COGNITIVE SCIENCE
A Multidisciplinary Journal
of
Anthropology, Artificial Intelligence, Education, Linguistics, Neuroscience, Philosophy, Psychology

Volume 29, Number 3   May/June 2005

Contents

SPECIAL ISSUE

**2004 Rumelhart Prize Special Issue Honoring John R. Anderson:
Theoretical Advances and Applications of Unified Computational Models**

The Cognitive Science Society Governing Board

Cognitive Science: A Multidisciplinary Journal is published six times a year by Lawrence Erlbaum Associates, Inc., 10 Industrial Avenue, Mahwah, NJ 07430–2262. Subscriptions for the 2005 volume are available only on a calendar-year basis.

Cognitive Science is the official journal of the Cognitive Science Society, Inc.

First published 2005 by Lawrence Erlbaum Associates, Inc

Published 2015 by Psychology Press
711 Third Avenue, New York, NY 10017

and by Psychology Press
27 Church Road, Hove, East Sussex, BN3 2FA

Psychology Press is an imprint of the Taylor & Francis Group, an informa business

This journal is indexed or abstracted in: *Math Rev, MLA Int Bibliography, Books Artic., Modern Language Literature (Complete Edition), Sociological Abstracts, PsychINFO, Psychol Abstracts* (1987–), *Comput Control Abstracts* (1987–), *Artificial Intelligence Abstracts, MCI Abstracts,* and *Ergonomics Abstracts.*

Copyright © 2005, Cognitive Science Society, Inc. All rights of reproduction in any form reserved.

ISBN 13: 978-0-8058-9426-4 (pbk)   ISSN 0364–0213

Cognitive Science 29 (2005) 307–311

Introduction to the 2004 Rumelhart Prize Special Issue Honoring John R. Anderson

This special issue honors the research and mentorship contributions of Dr. John R. Anderson, the 2004 David E. Rumelhart Prize recipient. This prize was instituted in 2001, funded by the Robert J. Glushko and Pamela Samuelson Foundation. The prize is awarded annually to an individual or collaborative team making a significant contemporary contribution to the formal analysis of human cognition. Mathematical modeling of human cognitive processes, formal analysis of language and other products of human cognitive activity, and computational analyses of human cognition using symbolic or nonsymbolic frameworks all fall within the scope of the award. In 2004 the prize selection committee consisted of Alan Collins, Robert Glushko, Mark Liberman, Anthony Marley, and James McClelland (chair). Perhaps best known for his contributions to connectionist or neural network models, Dr. Rumelhart also exploited symbolic models of human cognition, formal linguistic methods, and the formal tools of mathematics. Reflecting this diversity, the first three winners of the David E. Rumelhart Prize are individuals whose work lies within three of these four approaches. Past recipients are Geoffrey Hinton, a connectionist modeler, Richard M. Shiffrin, a mathematical psychologist, and Aravind Joshi, a formal and computational linguist. Anderson is the leading proponent of the symbolic modeling framework, thereby completing coverage of the four approaches.

Research Biography of John R. Anderson

John R. Anderson, Richard King Mellon Professor of Psychology and Computer Science at Carnegie Mellon University, is an exemplary recipient for a prize that is intended to honor "a significant contemporary contribution to the formal analysis of human cognition." For the last 3 decades, Anderson has been engaged in a vigorous research program with the goal of developing a computational theory of mind. Anderson's work is formulated within the symbol processing framework and has involved an integrated program of experimental work, mathematical analyses, computational modeling, and rigorous applications. His research has provided the field of cognitive psychology with comprehensive and integrated theories. Furthermore, it has had a real impact on educational practice in the classroom and on student achievement in learning mathematics.

Anderson's contributions have arisen across a career that consists of five distinct phases. Phase 1 began when he entered graduate school at Stanford at a time when cognitive psychology was incorporating computational techniques from artificial intelligence. During this period and immediately after his graduation from Stanford, he developed a number of simulation models of various aspects of human cognition such as free recall (Anderson & Bower, 1972). The human association memory (HAM) theory that he and Gordon Bower developed (Anderson & Bower, 1973) immediately attracted the attention of everyone then working in the field. The book played a major role in establishing propositional semantic networks as the basis for representation in memory and spreading activation through the links in such networks as the basis for retrieval of information from memory. It also provided an initial example of a research style that has become increasingly used in cognitive science—the creation of a comprehensive computer simulation capable of performing a range of cognitive tasks and the comparison of this model with a series of experiments addressing the phenomena within that range.

Dissatisfied with the limited scope of his early theory, Anderson undertook the work that has been the major focus of his career to date, the development of the Adaptive Control of Thought (ACT) theory (Anderson, 1976). ACT extended the HAM theory by combining production systems with semantic nets and the mechanism of spreading activation. The second phase of Anderson's career is associated with the initial development of ACT. The theory reached a significant level of maturity with the publication in 1983 of *The Architecture of Cognition* (Anderson, 1983), which is the most cited of his research monographs (having received over 2,000 citations in the ensuing years). At the time of publication, the ACT* model described in this book was the most integrated model of cognition that had then been crafted and tested. It has had a major impact on the theoretical development of the field and on the movement toward comprehensive and unified theories, incorporating separation of procedural and declarative knowledge and a series of mechanisms for production rule learning that became the focus of much subsequent research on the acquisition of cognitive skills. In his *Unified Theories of Cognition,* Alan Newell had this to say: "ACT* is, in my opinion, the first unified theory of cognition. It has pride of place. ... [It] provides a threshold of success which all other candidates ... must exceed" (Newell, p. 29).

Anderson then began a major program to test whether ACT* and its skill acquisition mechanisms actually provided an integrated and accurate account of learning. He started to apply the theory to development of intelligent tutoring systems; this defines the third phase of his research. This work grew from an initial emphasis on teaching the programming language LISP to a broader focus on high school mathematics (Anderson, Corbett, Koedinger, & Pelletier, 1995), responding to perceptions of a national crisis in mathematics education. These systems have been shown to enable students to reach target achievement levels in a third of the usual time and to improve student performance by a letter grade in real classrooms. Anderson guided this research to the point where a full high school curriculum was developed that was used in urban schools. Subsequently, a separate corporation was created to place the tutor in hundreds of schools, influencing tens of thousands of students. The tutor curriculum was recently recognized by the Department of Education as one of five "exemplary curricula" nationwide. Although Anderson does not participate in that company, he continues research to develop better tools for tracking individual student cognition, and this research continues to be informed by the ACT theory. His tutoring systems have established that it is possible to impact education with rigorous simulation of human cognition.

In the late 1980s, Anderson began work on what was to define the fourth phase of his research, which was an attempt to understand how the basic mechanisms of a cognitive architecture were adapted to the statistical structure of the environment. Anderson (1990) called this a rational analysis of cognition and applied it to the domains of human memory, categorization, causal inference, and problem solving. He used Bayesian statistics to derive optimal solutions to the problems posed by the environment and showed that human cognition approximated these solutions. Such optimization analysis and use of Bayesian techniques have become increasingly prevalent in cognitive science.

Subsequent to the rational analysis effort, Anderson has turned his full attention back to the ACT theory, defining the fifth and current phase of his career. With Christian Lebiere, he has developed the Adaptive Control of Thought–Rational (ACT–R) theory, which incorporates the insights from his work on rational analysis (Anderson & Lebiere, 1998). Reflecting the developments in computer technology and the techniques learned in the applications of ACT*, the ACT–R system was made available for general use. A growing and very active community of well over 100 researchers is now using it to model a wide range of issues in human cognition, including dual tasking, memory, language, scientific discovery, and game playing. It has become increasingly used to model dynamic tasks such as air traffic control, where it promises to have training implications equivalent to the mathematics tutors. Through the independent work of many researchers, the field of cognitive science is now seeing a single unified system applied to an unrivaled range of tasks. Much of Anderson's own work on the ACT–R has involved relating the theory to data from functional brain imaging (Anderson, Qin, Sohn, Stenger, & Carter, 2003).

In addition to his enormous volume of original work, Anderson has found the time to produce and revise two textbooks, one on cognitive psychology (Anderson, 2005) and the other on learning and memory (Anderson, 2000). The cognitive psychology textbook, now in its sixth edition, helped define the course of study that is modern introductory cognitive psychology. His more recent learning and memory textbook, now in its second edition, is widely regarded as reflecting the new synthesis that is occurring in that field in animal learning, cognitive psychology, and cognitive neuroscience.

Anderson has previously served as president of the Cognitive Science Society and has received a number of awards in recognition of his contributions. In 1978 he received the American Psychological Association's Early Career Award, in 1981 he was elected to membership in the Society of Experimental Psychologists, in 1994 he received APA's Distinguished Scientific Contribution Award, and in 1999 he was elected to both the National Academy of Sciences and the American Academy of Arts and Science. Currently, as a member of the National Academy, he is working toward bringing more rigorous science standards to educational research.

The Special Issue on Theoretical Advances and Applications of Unified Computational Models

This special issue of *Cognitive Science* features work by John Anderson's students, colleagues, and collaborators. The diversity of these articles attests to the fertility and generality of the ACT–R framework. It is a rare framework that can be employed for cognitive processes from driving simulations (Salvucci) and Web navigation (Pirolli) to sentence parsing (Lewis) and algebra (Anderson). It might be thought that a framework could only accommodate this diverse range of tasks by being a rather unconstrained general-purpose language. These contributions disprove this contention. ACT–R is powerfully grounded by rational considerations, and specific Bayesian formalisms lie at its core. Interestingly, ACT–R is not only constrained from above by considerations of rational use of probabilistic information, but is also constrained from below by recent neuroscience evidence (Anderson et al., 2003). The contributions by Lewis, Lovett, Pirolli, Salvucci, and Taatgen (this issue) show that a variety of detailed situations can be implemented in ACT–R, whereas Anderson's article (this issue) suggests how ACT–R, in turn, may be implemented in the brain.

ACT–R's combination of neural, computational and mathematical considerations impose strong constraints on the timing and control of cognitive processes, and impressively these constraints have been repeatedly shown to apply to human performance (for examples, see Lewis' constraints on sentence parsing, and Lovett's constraints on learning utility values of information sources). These articles attest not only to the broad applicability of ACT–R framework, but also to its genuine predictiveness. In short, the framework has proved to be a felicitous combination of a general and appropriately constrained language for expressing theories.

The articles reveal current lively areas of research for ACT–R specifically, and unified cognitive models more generally. One area of recent activity, featured prominently in the articles by Lovett, Salvucci, and Taatgen, is the need for mechanistic models of control processes. This focus is a sign of the maturity of cognitive science. In early computational models of cognition, it was considered sufficiently impressive to model performance on one task. Several of the articles in this special issue raise the bar for cognitive models, arguing that we need to know how a system behaves when confronted with several ongoing tasks. When the need to accomplish multiple tasks is acknowledged, mechanisms are needed to shift between tasks, foreground backgrounded processes, prioritize tasks, integrate and/or modularize processing, and monitor success and update plans accordingly. The authors tackle all of these hard issues with rigorous aplomb and take sizable steps toward grounding cognitive control in productions (Lovett, Taatgen, Salvucci) and brain structures (see Anderson's discussion of the role of the anterior cingulate in attentional control).

Another crucial aspect of human cognition that has been historically underrepresented is the surprising degree to which people can program themselves to strategically process their world. Failures of strategic programmability are prominent in cognitive science. Phenomena such as Stroop interference (see Lovett's article, this issue) are intriguing because they suggest that we are not in complete control of our cognitive processes. When instructed to name the color of a word's ink, participants are slower if the word is the name of a conflicting color than if it is a neutral word. Similarly, social psychologists have studied situations where people automatically tend to associate faces from other races with negative words at the same time that they associate faces from their own race with positive words, despite their stated desire not to engage in racial stereotypes and their explicit denial of the validity of these stereotypes. Demonstrations of these failures to have the thoughts we want to have are important because we generally believe we are in control of our own mental processes. However, it is equally important to reflect on the fact that these failures of control are only striking precisely because we know that we often *can* program ourselves with impressive levels of efficiency and reliability. Parking behavior is radically affected by presenting facts to oneself such as "The parking meters are not checked on Sundays" and "Snowplows come through this street at 3:00 a.m. when it snows." A single sentence of instructions to pay attention to color and not shape (Lovett), road conditions rather than incoming cell phone calls (Salvucci), or to use a specific method for solving algebraic problems (Anderson) can radically affect performance. Most computational models of cognition have ignored that people can be programmed by language to immediately alter their cognition. Several of these articles in this special issue (Anderson, Lovett, Salvucci, & Taatgen) embrace the unmet challenge of constructing cognitive devices that can be dynamically programmed by task needs, instructions, and changing strategies.

A final theme common to several of the manuscripts (most notably in the articles by Anderson, Pirolli, and Taatgen) is the fertility of applying rational and Bayesian principles to human cognition. The apparent wisdom of this approach appears on the surface to be cast in doubt by the large number of empirical demonstrations of suboptimal and irrational human decision making, including the research for which Daniel Kahneman was awarded a 2002 Nobel Prize in economics. Failures of rational decision making notwithstanding, enormous progress in many fields has been made with Bayesian modeling. This point is emphasized by the articles in this issue—they demonstrate surprising depth to rational analyses of memory, inference, search, and information access. Humans may not be built in the best way possible, but comparing our performance to formally devised gold standards of rationality has revealed both informative deviations from optimality and surprising cases where cognition apparently conforms to normative accounts.

Selected Bibliography

Anderson, J. R. (1976). *Language, memory, and thought.* Hillsdale, NJ: Lawrence Erlbaum Associates, Inc.

Anderson, J. R. (1983). *The architecture of cognition.* Cambridge, MA: Harvard University Press.

Anderson, J. R. (1990). *The adaptive character of thought.* Hillsdale, NJ: Lawrence Erlbaum Associates, Inc.

Anderson, J. R. (2000). *Learning and memory* (2nd ed.). New York: Wiley.

Anderson, J. R. (2005). *Cognitive psychology and its implications* (6th ed.). New York: Worth.

Anderson, J. R., & Bower, G. H. (1972). Recognition and retrieval processes in free recall. *Psychological Review, 79,* 97–123.

Anderson, J. R., & Bower, G. H. (1973). *Human associative memory.* Washington, DC: Winston.

Anderson, J. R., Corbett, A. T., Koedinger, K., & Pelletier, R. (1995). Cognitive tutors: Lessons learned. *Journal of the Learning Sciences, 4,* 167–207.

Anderson, J. R., & Lebiere, C. (1998). *The atomic components of thought.* Mahwah, NJ: Lawrence Erlbaum Associates, Inc.

Anderson, J. R., Qin, Y., Sohn, M-H., Stenger, V. A., & Carter, C. S. (2003). An information-processing model of the BOLD response in symbol manipulation tasks. *Psychonomic Bulletin & Review, 10,* 241–261.

Newell, A. (1990). *Unified theories of cognition.* Cambridge, MA: Harvard University Press.

Cognitive Science 29 (2005) 313–341

Human Symbol Manipulation Within an Integrated Cognitive Architecture

John R. Anderson

Psychology Department, Carnegie Mellon University

Received 28 April 2004; received in revised form 12 September 2004; accepted 20 September 2004

Abstract

This article describes the Adaptive Control of Thought–Rational (ACT–R) cognitive architecture (Anderson et al., 2004; Anderson & Lebiere, 1998) and its detailed application to the learning of algebraic symbol manipulation. The theory is applied to modeling the data from a study by Qin, Anderson, Silk, Stenger, & Carter (2004) in which children learn to solve linear equations and perfect their skills over a 6-day period. Functional MRI data show that: (a) a motor region tracks the output of equation solutions, (b) a prefrontal region tracks the retrieval of declarative information, (c) a parietal region tracks the transformation of mental representations of the equation, (d) an anterior cingulate region tracks the setting of goal information to control the information flow, and (e) a caudate region tracks the firing of productions in the ACT–R model. The article concludes with an architectural comparison of the competence children display in this task and the competence that monkeys have shown in tasks that require manipulations of sequences of elements.

Keywords: mathematics; cognitive architecture; education; learning; problem solving; comparative psychology; brain imaging

1. Introduction

Adaptive Control of Thought–Rational (ACT–R; Anderson et al., in press; Anderson & Lebiere, 1998) is most fundamentally a theory of central cognition. One function of this article is to present an overview of that theory. This article also presents an illustrative application of the theory to algebra equation solving. Algebra equation solving is a uniquely human cognitive activity and provides a relatively well-contained opportunity to address the question of what is unique about human cognition. I compare the requirements of this task with the requirements of other sequential tasks that nonhuman primates have been shown capable of performing. The article emerges with a tentative proposal for what is unique about human cognition, from the

Requests for reprints should be sent to John R. Anderson, Psychology Department, Carnegie Mellon University, Pittsburgh, PA 15213. E-mail: ja@cmu.edu

framework of the ACT–R theory. Such comparative analyses relate to issues of brain realization, and this article also describes the preliminary mapping of components of the ACT–R theory onto brain regions. This mapping has enabled the use of functional MRI (fMRI) data to inform theory development.

2. The ACT–R architecture

According to the ACT–R theory, cognition emerges through the interaction of a number of independent modules. Fig. 1 illustrates the modules relevant to algebra equation solving:

1. A visual module that might hold the representation of an equation such as $3x - 5 = 7$.
2. A problem state module (sometimes called an imaginal module) that holds a current mental representation of the problem. For instance, the student might have converted the original equation into $3x = 12$.
3. A control module (sometimes called a goal module) that keeps track of one's current intentions in solving the problem—for instance, one might be trying to apply the unwind strategy described later.
4. A declarative module that retrieves critical information from declarative memory such as that $7 + 5 = 12$.
5. A manual module that programs the output such as $x = 4$.

Each of these modules is capable of massively parallel computation to achieve its objectives. For instance, the visual module is processing the entire visual field and the declarative module searches through large databases. However, each of these modules suffers a serial bottleneck such that only a little information can be put into a buffer associated with the module—a single object is perceived, a single problem state represented, a single control state

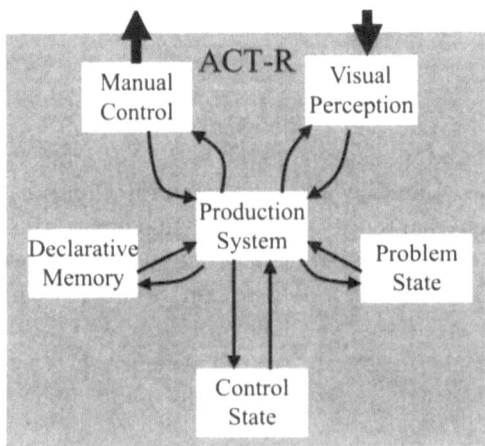

Fig. 1. The interconnections among modules in ACT–R 5.0.

maintained, a single fact retrieved, or a single program for hand movement executed. Formally, each buffer can only hold what is called a *chunk* in ACT–R, which is a structured unit bundling a small amount of information. ACT–R does not have a formal concept of a working memory, but the current state of the buffers constitutes an effective working memory. Indeed, there is considerable similarity between these buffers and Baddeley's (1986) working memory "slave" systems.

Communication among these modules is achieved via a procedural module (production system in Fig. 1). The procedural module can respond to information in the buffers of other modules and put information into these buffers. The response tendencies of the central procedural module are represented in ACT–R by production rules. For instance, the following might be a production rule for transforming an equation:

> *IF* the goal is to solve the equation
>   and the equation is of the form Expression – Number1 = Number2
>   and Number1 + Number2 is Number3 has been retrieved
> *THEN* transform the equation to Expression = Number3

This production responds when the control chunk encodes the intention to solve an equation, as shown in the first line; when the problem state chunk represents an equation of the appropriate type—second line—for example, $3(x-2)-4=5$; when a chunk encoding an arithmetic fact has been retrieved from memory—see third line—in this case $4+5=9$; and appropriately changes the problem representation chunk—see fourth line—in this case to $3(x-2)=9$.

The procedural module is also capable of massive parallelism in sorting out which of its many competing rules to fire, but like the other modules it has a serial bottleneck in that it can only fire a single rule at a time. Because it is responsible for communication among the other modules, the production system comprises the central bottleneck (Pashler, 1994) in the ACT–R theory. Therefore, cognition can be slowed when there are simultaneous demands to process information in distinct modules. As already noted, the other modules themselves can also be bottlenecks. All of the bottlenecks are in the communication among modules; within modules things are massively parallel. (Fig. 3, later in the paper, illustrates in some considerable detail how this parallelism and seriality mix.) Documenting the accuracy of this characterization of human cognition has been one of the preoccupations of research on ACT–R (for instance, Anderson, Taatgen, & Byrne, 2004).

In addition to the overall flow of control, major concerns of ACT–R involve the "internal components" (declarative memory, production memory, control state, and problem state). With respect to peripheral modules (and ACT–R has more than just the visual and manual modules represented here) we have been content to implement approximations that capture the major results documented in the literature. Indeed, much of ACT–R's perceptual–motor system is a reimplementation of Executive-Process Interactive Control's (EPIC; Meyer & Kieras, 1997) perceptual–motor system. Following EPIC's lead we have found that we cannot understand central cognition unless we have reasonably accurate models of its interface with the external world. For a substantial fraction of the ACT–R community, particularly those concerned with human–computer interaction issues, this perceptual–motor system is critical.

The central declarative memory and the procedural production system have substantial similarities. Selection among different declarative memories proceeds by means of computations on continuously valued activation quantities, and selection among productions proceeds by means of continuously valued utility computations. These computations give ACT–R many of the graded properties considered virtues of connectionist systems and avoid the sharp edges associated with symbolic systems. Indeed, there was a connectionist implementation of ACT–R (Lebiere & Anderson, 1993). However, the theory also has an important symbolic level that plays a critical role in accounting for such things as the acquisition of competence in algebra. The distinction between the symbolic level of facts and productions and the subsymbolic level of activations and utilities is critical to the ACT–R theory.

Until recently, the problem state and the control state were merged into a single goal system. There have been a number of developments to improve ACT–R's goal system (Altmann & Trafton, 2002; Anderson & Douglass, 2001), and this is another development. There were two reasons for choosing to separate control state (*goal buffer*) and problem state knowledge (*imaginal buffer*). First (and this was the source of the idea to separate the two aspects), our imaging data indicated that the parietal region of the brain reflected changes to problem-state information, but the anterior cingulate reflected control-state changes. Later, this article elaborates on the neural basis for this distinction. Second, the distinction offered a solution to a number of nagging problems we had with the existing system that merged the two types of knowledge. One problem was that our goal chunks often seemed too large, violating the spirit of the claim that chunks were supposed to only contain a little information. This is because they contained both problem-state information and control-state information, which could both involve a number of elements.[1] Also, the control information was getting in the way of storing useful information about the problem solution in declarative memory. For instance, arithmetic facts such as $3 + 4 = 7$ might represent the outcome of a counting process or of an effort to comprehend a sentence. Because the control information was separate and would be different in the case of these two sources for the same arithmetic fact, we effectively were creating parallel memories storing the same essential information. Now, with control and problem state separated, the differences between the counting and comprehension can be represented in different control chunks, whereas the common result would be represented identically in a single problem solution chunk. By factoring control information away (in what we are now calling the *goal module*), one can accumulate abstract memories of the information achieved in the problem state.

3. Algebra equation manipulation

With this brief overview of the ACT–R theory, let us turn to algebra equation solving, which is a domain that offers special opportunities for understanding the nature of human intelligence. Although there are now a number of demonstrations of basic arithmetic competence in other primates, it would be generally conceded that algebra is a uniquely human capability. Algebraic expressions and the operations that can be performed on them represent a domain of substantial cognitive complexity, but unlike many human accomplishments (such as natural language) it is a domain that can be tractably characterized and studied. In the first year of high school algebra with an investment of typically less than 200 hr, suitably prepared students can learn to take a sequence of symbols such as

$$4 = x *[(x + 8) + 7] \qquad (1)$$

and rewrite it into the form

$$x^2 + 15x - 4 = 0 \qquad (2)$$

in preparation for using the quadratic formula, which they can also apply. Interestingly, students come to prefer writing this out directly without writing any intermediate expressions—despite the urging of some mathematics teachers not to do so. Children have such a facility at manipulating these representations in their heads that it is easier to do the operations mentally than to write out intermediate expressions.

Before continuing this discussion, it is important to make a couple of caveats. First, I am not implying that the ability to engage in such symbol manipulation is the most important part of first-year high school algebra, nor that students spend the majority of their 200 hr mastering this. The goal of the algebra course is to relate multiple representations of mathematical relations (including graphical and verbal) to enable flexible problem solving (Koedinger, Anderson, Hadley, & Mark, 1997). Just being able to engage in such manipulations would be a rather useless skill unless students could relate such manipulations to other things, especially real-world problems. Nonetheless, even though such algebraic manipulations are a small part of the complete picture, they already establish a high level of complexity to human cognition—a level of complexity that is all the more remarkable given that it is mastered in such a brief period of time.

Second, being able to achieve this competence depends critically on what has already been established in earlier grades—in particular, knowledge of arithmetic facts, fractional representations, and how to parse expressions such as the previous example. Students struggle if they arrive in algebra without these prerequisites. Thus, the modeling task in this article is to account for the acquisition and performance of algebraic transformations, assuming the background of such knowledge.

As the previous example illustrates, algebra manipulation is basically a string manipulation task in which one string of symbols is transformed into another. The final section of this article considers what might be uniquely human about this string manipulation task versus other sequential skills that nonhuman primates can do.

The experiment to be modeled in detail looks at the learning of a particularly reduced version of algebra symbol manipulation—solving of simple linear equations—converting expressions such as

$$3x - 5 = 7 \qquad (3)$$

into $x = 4$. $\qquad (4)$

These sorts of equations can be solved by what has been called the *unwind* strategy. Such equations have a number on one side and an expression with a single occurrence of the variable on the other. The variable can be isolated by inverting each operation (in the previous example the "−5" is eliminated by adding 5, and the "3*" by dividing by 3)—peeling away the layers of the expression until the variable is exposed. Many equations do not immediately start out in this form but can be simplified so that they are in the appropriate form. The model in this article assumes that the equations are already in a form to which the unwind strategy can immediately apply. The earliest instruction on equation solving tends to focus on such problems and teaches students the justifications for these transformations as well as providing practice on how to perform them, although students are typically not taught to think of this as a general unwind strategy but rather as a series of more specific operations. We have found, however, that beginning students are quite capable of understanding the general unwind principle and its justification and can use it with its full generality.

In the experiment to be modeled in detail (Qin et al., 2004) 10 students ages 11 to 14 spent 6 days practicing solving such equations. The first day (Day 0) they were given private tutoring on this class of equations, using the unwind principle, and practiced paper and pencil solutions of such problems with a private human tutor. On the remaining 5 days they practiced on a computer the solution of three classes of equations:

$$\text{0-step: e.g., } 1x + 0 = 4 \tag{5}$$

$$\text{1-step: e.g., } 3x + 0 = 12, \ 1x + 8 = 12 \tag{6}$$

$$\text{2-step: e.g., } 7x + 1 = 29 \tag{7}$$

Each day they went through 10 computer-administered blocks of such equations. Each block consisted of 16 trials with four instances of the four possible types of equations (there are two subtypes for the 1-step equations). Fig. 2 presents their latency and the predictions of a model that will now be described.

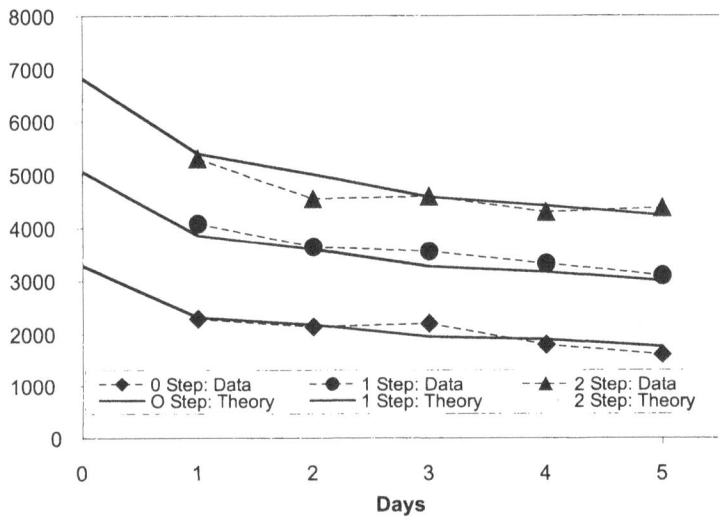

Fig. 2. Mean solution times (and predictions of the ACT–R model) for the three types of equations as a function of delay. Although the data were not collected, the predicted times are presented for the practice session of the experiment (Day 0).

4. The ACT–R model

The ACT–R model begins with a set of declarative instructions, given in Table 1, which encode the unwind strategy. To illustrate how these instructions[2] apply to example equations, first consider a simple 0-step equation such as:

$$1*x + 0 = 2 \tag{8}$$

These instructions imply a sequence of operations that can be summarized:

Instruction 1a: Create image " = 2"
Instruction 2b: Unwind right "$1*x + 0$"
Instruction 3a: Focus on "$1*x$" and unwind it
Instruction 2c: Unwind left "$1*x$"
Instruction 4a: Focus on "x" and unwind it
Instruction 2a: The answer is 2

While for a 2-step equation such as

$$7*x + 3 = 38 \tag{9}$$

they imply a sequence of operations that can be summarized:

Instruction 1a: Create image " = 38"
Instruction 2b: Unwind right "$7*x + 3$"
Instruction 3b: Change image to " = 38 – 3" and then to " = 35" and focus on "$7*x$" and unwind it.
Instruction 2c: Unwind left "$7*x$"
Instruction 4b: Change image to " = 35/7", and then to " = 5" and focus on x and unwind it.
Instruction 2a: The answer is 5

Fig. 2 shows that ACT–R is able to reproduce the speedup seen in the participants. The key to understanding this speedup in the ACT–R model is to understand how the previous instructions were interpreted. These instructions are encoded as declarative structures, and ACT–R has general interpretative productions for converting these instructions to behavior. For instance, there is a production rule that retrieves the next step of an instruction:

IF one has retrieved an instruction for achieving a goal
THEN retrieve the first step of that instruction

There are also productions for retrieving particular arithmetic facts such as

IF one is evaluating the expression "a operator b"
THEN try to retrieve a fact of the form "a operator b = ?"

Using such general instruction-following productions is laborious and accounts for the slow initial performance of the task.

Although multiple types of learning are occurring in this experiment, it is mainly production compilation that is accounting for the speedup (see Taatgen, 2005, this issue; Taatgen & Anderson, 2002). This is a process by which new production rules are learned that collapse what was originally done by multiple production rules. In this situation the initial instruction-following productions are compiled over time to produce productions to embody procedures that efficiently solve equations. For instance, the following production rule is acquired:

IF the goal is to unwind an expression
   and the expression is of the form "subexpression + 0"
THEN focus on the subexpression

Fig. 3a illustrates a typical trial at the beginning of Day 1, and Fig. 3b illustrates a typical trial at the end of Day 5. In both cases the model is solving the 2-step equation, $7*x + 3 = 38$. The figure illustrates when the various modules were active during the solution of the equation and what they were doing. The Day 1 trial (Fig. 3a) takes 6.1 sec and the Day 5 trial (Fig. 3b) takes 4.1 sec. However, these do not reflect the extremes of the learning curve according to ACT–R. The very first trial on Day 0 takes 8.4 sec in the model. With an infinite amount of practice, the model would take 1.7 sec during which it would only read the equation and type the answer, having compiled the answer into production rules for that problem. Still, the contrast between parts a and b of Fig. 3 gives a sense for what is happening over the course of learning. It is worth emphasizing a number of general features of the activity in the figure before discussing the detail of what is happening in individual buffers:

Multiple modules can be active simultaneously. For instance, early on in Fig. 3 there is a point where the goal module is noting that it is implementing the unwind strategy, an image of the right-hand side of the equation (" = 38") is being encoded in the imaginal buffer, the next step in the unwind strategy is being retrieved, and the visual system is encoding the left-hand side of the equation. Certain of these activities tend to be on the critical path because they are taking longer than the other processes, and further processing has to wait for them to complete. In these cases, the times of the other operations have no effect on total time. For instance, often the visual encoding of the equation is holding up other operations and the durations of these other operations do not matter.

Table 1
English rendition of instructions given to ACT–R model for equation solving

---

1. To solve an equation, encode it and
   a. If the right side is a number, then imagine that number as the result, and focus on the left side and unwind it.
   b. If the left side is a number, then imagine that number as the result, and focus on the right side and unwind it.
2. To unwind an expression
   a. If the expression is the variable, then the result is the answer.
   b. If a number is on the right unwind-right.
   c. If a number is on the left unwind-left.
3. To unwind-right, encode the expression (of the form "subexpression operator number") and
   a. If the operator is + or – and the number is 0, then focus on the subexpression and unwind it.
   b. Otherwise invert the operator, imagine it as the operator in the result, imagine the number of the expression as the second argument in the result, evaluate the result, and then focus on the subexpression and unwind it.
4. To unwind-left encode the expression (of the form "number operator subexpression") and
   a. If the operator is * and number 1 then focus on the subexpression and unwind it.
   b. Otherwise check that the operator is symmetric, invert the operator, imagine it as the operator in the result, imagine the number as the second argument in the result, evaluate the result, and then focus on the subexpression and unwind it.

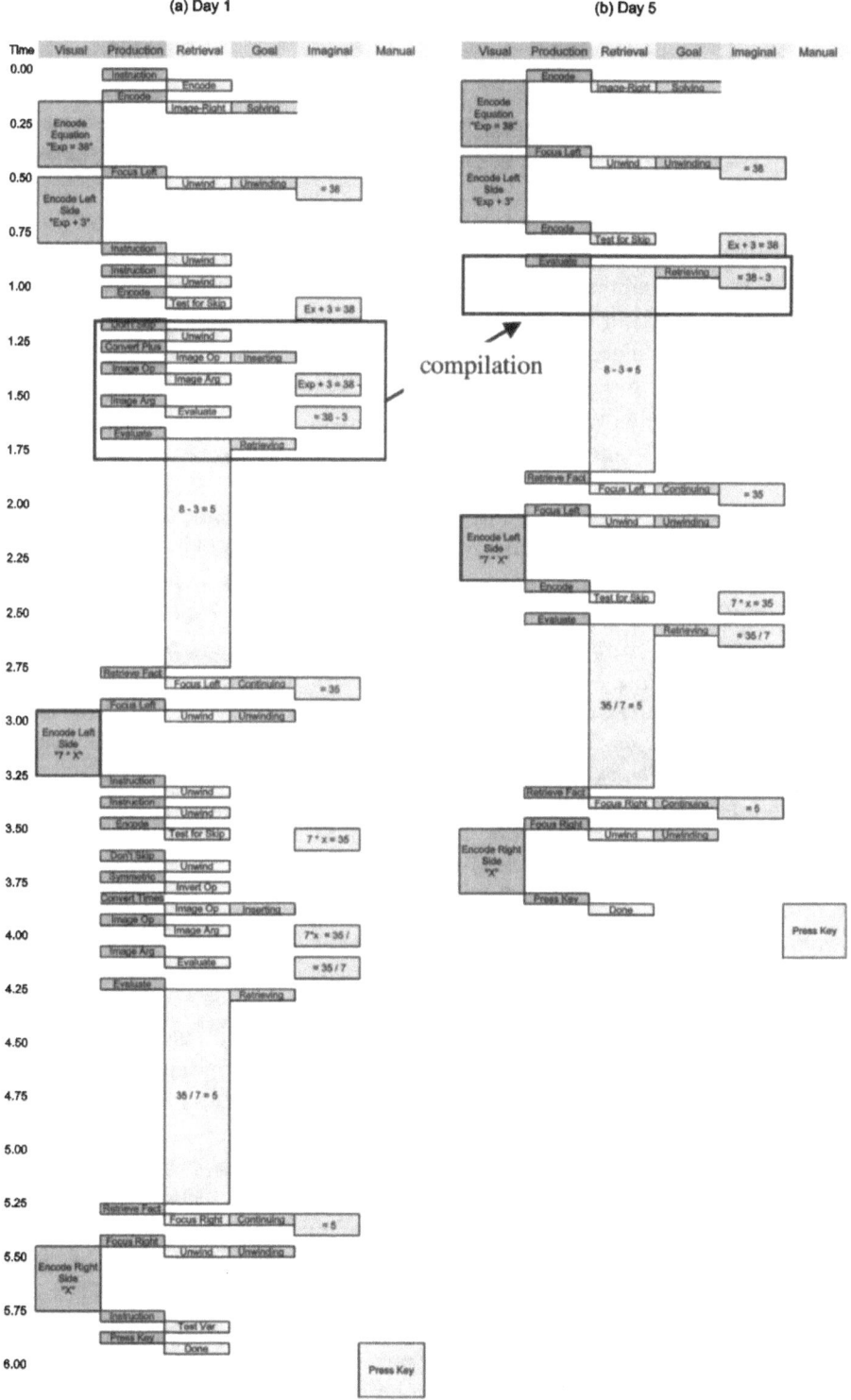

Fig. 3. Comparison of the module activity in ACT–R during the solution of a two-step equation on Day 1 (a) with a two-step equation on Day 5 (b). In both cases the equation being solved is $7*x + 3 = 38$.

Much of the speedup in processing is driven by collapsing multiple steps into single steps. A particularly dramatic instance of this is noted in Fig. 3 where five production firings, five retrievals, two control settings of the goal, and two imaginal transformations are compressed into one production, one retrieval, one control setting, and one imaginal transformation.[3] Production compilation can compress these internal operations without limit. What it cannot collapse are the external operations such as visual encodings or manual operations. These external operations define the bounds of the compilation. Although the example in Fig. 3 shows multiple productions being collapsed, the actual learning process proceeds slowly in ACT–R and takes all 5 days to achieve the transformation in Fig. 3. Given enough practice ACT–R would collapse all equation solving simply into a series of visual encodings and manual operations, and there would be no effect of equation complexity (nor any real thought occurring). However, to do so ACT–R would have to essentially build into production rules the capacity to recognize each possible equation and produce its solution. The combinatorics of this are so overwhelming (so many different possible equations) that it would never happen in the normal course of learning to solve equations.

A second, lesser source of speedup is the reduction of retrieval times. This reflects an increase in the base-level activation of the facts used in this experiment and as such it is an example of subsymbolic activation learning. This subsymbolic learning is a relatively minor contributor to the learning in Fig. 2 for two reasons. First, the basic instructions get used over and over again and are already strongly encoded during Day 0, and there is not that much room for further speedup. Second, the arithmetic facts do not repeat very often over the course of the experiment and are getting little practice. In other situations, subsymbolic activation processes can be a major player in performance. However, over the period of time studied in this experiment, the major learning is happening at the symbolic level in terms of creating new production rules.

Now let us consider what is happening in each of the modules as the model goes from Day 1 to Day 5:

1. Visual: On both days four encoding operations take place, which each take 300 msec. Each encoding has the resolution to pick up two terms in the expression. Therefore, the first encodes $Exp = 38$, where Exp denotes what cannot be analyzed. The second analyzes this into $Exp + 3$, the third into $5 * Exp$, and the final encodes the x.

2. Procedural: The number of productions fired reduces substantially over the 5 days—from 26 to 11 in the example in Fig. 3. This reflects the compilation of productions into ones that do more work. Many of the productions, even on Day 1, were compiled from the originals that were used on Day 0. Each production takes 50 msec according to the ACT–R theory.[4]

3. Retrieval: Most of the retrievals in Fig. 3 involve retrieval of steps of instruction, and these decrease dramatically from 24 to 9 in the specific example. These instructions are illustrated as taking about 50 msec, but this is only approximate. There are also two long retrievals of arithmetic facts. As noted earlier, all of these retrievals are speeding up with time, but the speedup effect is most apparent for the arithmetic facts.

4. Goal: The goal holds information about control state. Different points in the problem solving can have identical patterns in the other buffers, and it is the responsibility of the goal buffer to keep track of what to do next. As I elaborate later, the major control issue is keeping track of when it is time to retrieve and when it is time to unwind. The number of control settings

changes relatively little over the course of the experiment, decreasing from 10 to 8 in the specific example.

5. Imaginal: The imaginal, or problem state, buffer holds a partial representation of the equation. The number of changes to this buffer shows a small drop in Fig. 3 from nine to seven. The reduction reflects cases when two imaginal operations are collapsed into a single one.

6. Manual: The manual programming does not change over the course of the experiment. A single final finger press is needed that takes 250 msec to program and execute.

5. Use of brain imaging to provide converging data

The complexity of the picture in Fig. 3 provides a striking contrast to the simplicity of the data in Fig. 2. It is hard to justify all those boxes and assumptions on the basis of three simple learning curves. Reflecting this, past theories that we have developed (e.g., Anderson, Reder, & Lebiere, 1996) have often been cast much simpler, not because we thought things were that simple but because we were keenly aware of the assumptions-to-data ratio. However, in honest moments with ourselves we knew more was going on. Indeed, Fig. 3 probably underrepresents the true complexity. Although for many purposes ignoring this complexity is fine, it left us with a picture of mathematics learning that may have failed some of our theory-based efforts to improve mathematics learning. It is better to have a complete theory and determine which details are not relevant and can be ignored, rather than simply never considering the details in the first place.

Brain-imaging data allow us to track these individual components in more detail. Although it still does not provide all of the converging evidence one would want, it goes a long way to justifying the detail in Fig. 3. The study reported in Fig. 2 was actually performed in an fMRI scanner on Days 1 and 5. The trials took 21.6 sec on all days, to facilitate analysis on the scanner days. On the scanner days, an image of much of the brain was taken each 1.2 sec. During the first 1.2 sec, children looked at a fixation point. Then they had up to 10 scans or 12 sec to complete solving the equation, and they pressed a key giving an answer as soon as they had solved the equation. These 10 scans were followed by an additional 7 scans or 8.4 sec to let the hemodynamic response to the equation go down to baseline. Students gave their answer in a data glove in which they pressed one of the 5 fingers on their right hand to indicate an answer of 1 to 5 (all problems had these numbers as answers).

5.1. Regions of interest

We have now completed a large number of fMRI studies of many aspects of higher level cognition (Anderson, Qin, Sohn, Stenger, & Carter, 2003; Anderson, Qin, Stenger, & Carter, 2004; Qin et al., 2003; Sohn, Goode, Stenger, Carter, & Anderson, 2003; Sohn et al., in press), and based on the patterns over these experiments we have made the following associations between a number of brain regions and modules in ACT–R. In this article we are concerned with five brain regions and their ACT–R associations:

1. Caudate (procedural): Centered at Talairach coordinates $x = -5$, $y = 9$, $z = 2$, this is a subcortical structure.

2. Prefrontal (retrieval): Centered at $x = -40$, $y = 21$, $z = 21$, this includes parts of Brodmann Areas 45 and 46 around the inferior frontal sulcus.
3. Anterior cingulate (goal): Centered at $x = -5$, $y = 10$, $z = 38$, this includes parts of Brodmann Areas 24 and 32.
4. Parietal (problem state or imaginal): Centered at $x = -23$, $y = -64$, $z = 34$, this includes parts of Brodmann Areas 39 and 40 at the border of the intraparietal sulcus.
5. Motor (manual): Centered at $x = -37$, $y = -25$, $z = 47$, this includes parts of Brodmann Areas 3 and 4 at the central sulcus.

It is important to emphasize that we had these regions defined and associated with ACT–R modules before performing this experiment. Thus, the research we are reporting here contrasts with the more typical practice of performing exploratory analyses to find what regions give significant changes in activation and trying to interpret their significance after the fact. As such this confirmatory approach is not subject to the issue of trying to correct for false alarms that haunts the exploratory approach.

It is worth briefly noting the past reports in which we identified these regions and associated them with the particular modules in the ACT–R theory. The original publication in the series (Anderson et al., 2003) was focused on looking for the brain correlates of the Anderson, et al. (1996) ACT–R model for algebra equation solving in adults. The two experiments in that article performed exploratory analyses that converged on regions close to the prefrontal, parietal, and manual regions defined previously. Based on that study, we defined these previously mentioned regions and conducted a number of studies focused on verifying their properties. Across these studies we maintained the exact same Talairach definition of these regions. The studies by Anderson et al. (2004), Anderson et al. (in press), & Qin et al. (2003) were focused on better separating of parietal and prefrontal activities (which are often highly correlated) and confirming that the prefrontal was more associated with retrieval and the parietal was more associated with representational changes. That research also showed that these regions responded to the number of retrievals and representational changes and not to the duration of time that the retrieval products or representations were held during the problem solving. Interestingly, we found that the motor region was involved in rehearsal of the results to bridge delay periods, not the parietal or prefrontal regions. The research by Sohn et al. (2003) and Sohn et al. (2005) used the fan effect to confirm that the prefrontal and not the parietal region responded to time to perform an individual retrieval.

Although all of these published articles only reported on the prefrontal, parietal, and motor regions, we did collect data on predefined anterior cingulate cortex (ACC) and caudate regions in all of these studies. Our interest in the ACC was forced by the fact that exploratory studies kept revealing that it showed strong effects from the experimental manipulations. The caudate region of the basal ganglia area only sometimes showed significant effects in the exploratory analysis, and we will see it suffers from a rather poor signal-to-noise ratio. However, our interest in it was driven by our association of the basal ganglia with the production system (Anderson et al., in press) and by other published reports associating the basal ganglia with procedural memory (Ashby & Waldron, 2000; Hikosaka eti al., 1999; Poldrack, Prabakharan, Seger, & Gabrieli, 1999; Saint-Cyr, Taylor, & Lang, 1988).

Although we have our own independent evidence for the associations of these regions with the ascribed ACT–R functions, it certainly is the case that the ascriptions are consistent with other ideas in the literature. As noted previously, our interest in the caudate was basically driven by other research reports. Others (Dehaene, Piazza, Pinel, & Cohen, 2003; Reichle, Carpenter, & Just, 2000) have found a parietal region that reflects imagery and visual representation, and a number of researchers (Buckner, Kelley, & Petersen, 1999; Cabeza, Dolcos, Graham, & Nyberg, 2002; Donaldson, Petersen, Ollinger, & Buckner, 2001; Fletcher & Henson, 2001; Lepage, Ghaffar, Nyberg, & Tulving, 2000; Wagner, Maril, Bjork, & Schacter, 2001; Wagner, Paré Blagoev, Clark, & Poldrack, 2001) have found a strong memory response in the vicinity of our prefrontal region.

The situation in the literature with respect to the ACC is complex. A number of theorists have postulated that it is involved in controlling cognition, much as is being proposed here. For instance, Posner & Dehaene (1994) have described the ACC as "involved in the attentional recruitment and control of brain areas to perform complex tasks" (p. 76). D'Esposito et al. (1995) have identified it with Baddeley's (1986) central executive and Posner & DiGirolamo (1998) have related it to Norman & Shallice's (1986) SAS. However, there are other theories of the ACC. One theory relates it to error detection. This is supported by the error-related negativity in event-related potentials that has been observed when errors are made in speeded response tasks (e.g., Falkenstein, Hohnsbein, & Hoorman, 1995). Dehaene, Posner, and Tucker (1994) were able to localize the error-related negativity as residing within the ACC. However, ACC activity occurs in many more situations than just when there are errors, and another interpretation of its activity is that it is just a reflection of task difficulty as indexed by errors or reaction time (Paus, Koski, Caramanos, & Westbury. 1998). On the other hand, it does not always respond to task-difficulty factors that affect latency (we will see in our experiment that it reflects number of transformations but not practice). Carter, et al. (2000) argued that the real function of the ACC is monitoring for conflict among potential responses and that other regions of the cortex actually respond to the conflict once detected. MacDonald, Cohen, Stenger, and Carter (2000) found that in a Stroop task, when participants are warned that it will be a difficult color trial, there is greater activation in the prefrontal region in preparation for the task. In contrast, when the actual Stroop task is presented the ACC responds to a difficult color trial. Thus, they argue that, unlike the Posner and Dehaene proposal, the prefrontal cortex, and not the ACC, is responsible for control and that the ACC, rather, monitors for conflict, such as that which occurs in the Stroop task. This conflict is often interpreted as conflict among competing responses, and this interpretation is applicable to the Stroop task. However, in our more complex tasks ACC activity reflects transformations and retrievals that do not involve any overt responses or competitions among responses. Sohn, Albert, Jung, Carter, & Anderson (2004) reported a study in which the ACC is clearly involved in controlling attention and not just monitoring conflict. Unlike the MacDonald et al. (2000) experiment, it is highly active in preparation for an upcoming cognitive task, and its activation varies with the anticipated difficulty of that task. At the end of this article we elaborate on the function for the ACC in the ACT–R model.

We should also comment on the restriction of these regions to the left hemisphere. In the case of the motor region this restriction is obvious because participants are responding with their right hand. We and the other researchers we mentioned have found stronger responses in the left parietal and left prefrontal in these kinds of symbolic tasks. The restriction to the left caudate and left ACC is largely done for consistency, but more often than not the response is stronger in the left region. This is somewhat surprising in the case of the ACC because the left and right ACC regions are adjacent to each other.

Given that these regions reflect their ascribed function so well, one is tempted to assume that the function is actually performed in that cortical region. Although this is a plausible inference that many make, it is not necessary to the logic of our approach. We only require that we have a brain region whose activity reliably reflects a particular information processing function. Even if we assume that the function is performed in that region, there is no reason to suppose that its activity will only reflect that function. Nonetheless, we have been fortunate over the series of studies that we have performed that the regions seem to be rather pure indicators of their as-cribed functions.

Finally, there is no claim that the ascribed function is restricted to these regions. With re-spect to retrieval we suspect there was a similar response in the hippocampus, but our scanning parameters in this experiment did not include the hippocampus, and it appears not to give as strong a signal when we do. With respect to the control, it is almost certain that dorsolateral, prefrontal structures play a role in control as well as the ACC. With respect to the caudate, we would expect to find a similar response in other structures connected to the basal ganglia, par-ticularly the dorsal thalamus (and indeed we often do). Elsewhere (Anderson et al., 2004) we have reviewed proposals (Amos, 2000; Frank, Loughry, & O'Reilly, 2000; Houk & Wise, 1995; Wise, Murray, & Gerfen, 1996) that the basal ganglia perform functions similar to those that we ascribe to the production system. Finally, note that our list of five regions does not con-tain a region that corresponds to the visual module. This is because our scanning parameters also did not include the relevant visual regions. Other studies have found the expected re-sponses in the visual cortex with visual presentation and the auditory cortex with auditory pre-sentation (Sohn et al., in press).

5.2. Predicting the BOLD response

We have developed a methodology for relating the profile of activity in modules such as those in Fig. 3 to blood-oxygen-level-dependent (BOLD) responses from the brain regions that correspond to these modules. Fig. 4 illustrates the proportion of time that a particular module was active at various points during a trial on Day 1 (Part a) and Day 5 (Part b) for the two-step equations. These numbers would be directly obtainable from Fig. 3, except that Fig. 4 reflects the average engagement over the whole day not just at the beginning of Day 1 (Fig. 3a) and the end of Day 5 (Fig. 3b). The basic model we have developed of the BOLD response claims that while a module is engaged, it is producing a hemodynamic response in the corresponding re-gion. We have adopted the standard gamma function that other researchers (e.g., Boyton, Engel, Glover, & Heeger, 1996; Cohen, 1997; Dale & Buckner, 1997; Glover, 1999) have used for the BOLD response. If the module is engaged it will produce a BOLD response t time units later according to the function:

$$B(t) = m \left(\frac{t}{s} \right)^a e^{-(t/s)}$$

where m governs the magnitude, s scales the time, and the exponent a determines the shape of the BOLD response such that with larger a the function rises and falls more steeply. The time to

a)

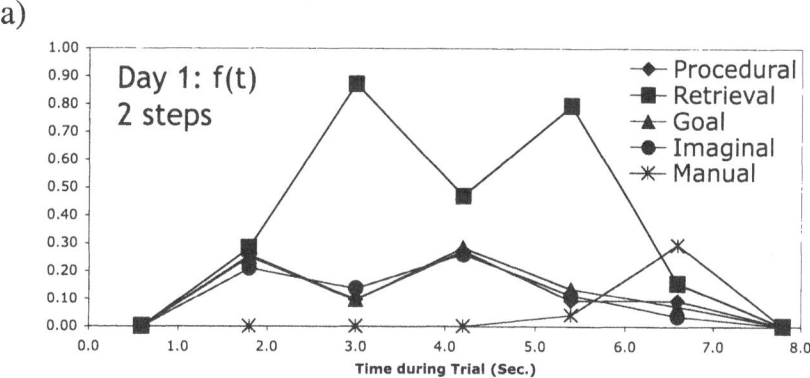

b)

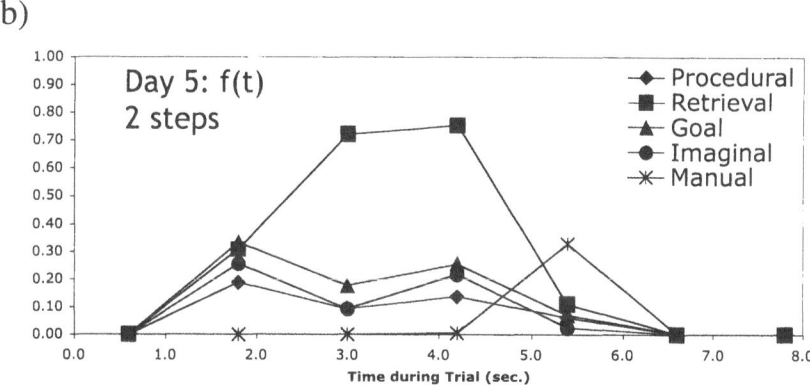

Fig. 4. The degree of engagement of the various modules during a trial on Day 1 (part a) and Day 5 (part b).

peak for the BOLD response is $a*s$, and the magnitude area under the curve is $m*s*\Gamma(a)$ where Γ is the gamma function, $[\Gamma(a) = (a - 1)!]$. Fig. 5 illustrates the effect of different choices of time scale and exponent on the shape of the BOLD response. To facilitate comparison, the magnitude parameters for the curves in this figure have been set so that the maximum response is 1 for all functions. As can be observed, a larger a produces a quicker rise and fall, whereas a larger s stretches the duration of the BOLD response.

The BOLD response accumulates whenever the region is engaged. Thus, if $f(t)$ is an engagement function giving the probability that the region is engaged at time t, then the cumulative BOLD response can be obtained by convolving the two functions:

$$CB(t) = \int_{0}^{t} f(x)B(t - x)dx$$

This is the observed BOLD response. Its area is proportional to the total time that the region is engaged. Thus, if a module is active for T sec, then the area under the BOLD response is $T*m*s*\Gamma(a)$.

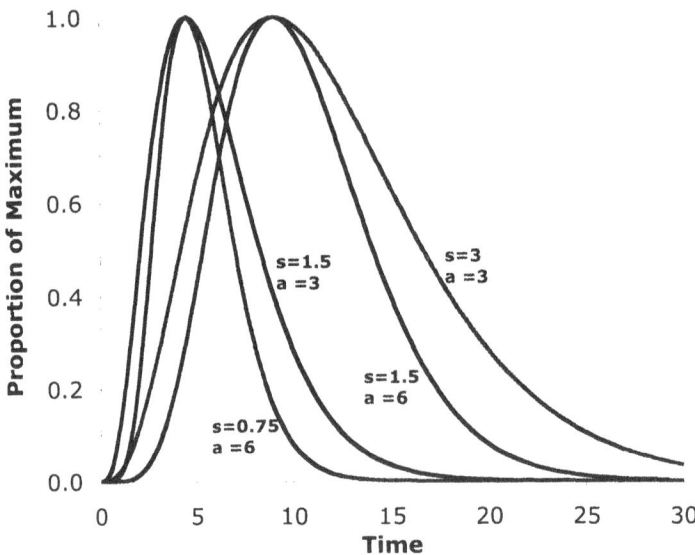

Fig. 5. An illustration of the impact of different choices of the exponent (a) and time scale on the shape of the hemodynamic function. To facilitate comparison, the magnitude parameter (m) has been scaled so that all of these functions have a maximum of 1.0.

In summary, a model for the time course (Fig. 3) of this task yields engagement functions $f(t)$ such as those in Fig. 4. By convolving the engagement functions with the BOLD function one can obtain predictions for the BOLD response in the regions associated with the modules. Most of the parameters of this model are set according to prior values established for ACT–R, but fitting the latency in Fig. 2 did require estimating parameters for the time to encode the equation and the duration of the retrievals. Having now committed to the time course of each module, predictions immediately follow for the time course of the cumulative BOLD response. The exact height and shape of the BOLD response depends on the magnitude (m), the scale (s), and the exponent (a) for the region that corresponds to that module. However, the strong parameter-free prediction is that the relative areas under the BOLD responses in two conditions for a region will reflect the relative amounts of time this region is engaged in these two conditions. Thus, the BOLD response provides a direct check on assumptions about the amount of time various modules are engaged in doing a task.

Table 2
Parameters estimated and fits to the BOLD response

	Motor/ Manual	Prefrontal/ Retrieval	Parietal/ Imaginal	Cingulate/ Goal	Caudate/ Procedural
Magnitude (m)	0.531	0.073	0.231	0.258	0.207
Exponent (a)	3	3	3	3	3
Scale (s)	1.241	1.545	1.645	1.590	1.230
Correlation	.975	.963	.969	.981	.975
Chi-square (105 df)	88.93	82.60	95.21	123.27	81.03

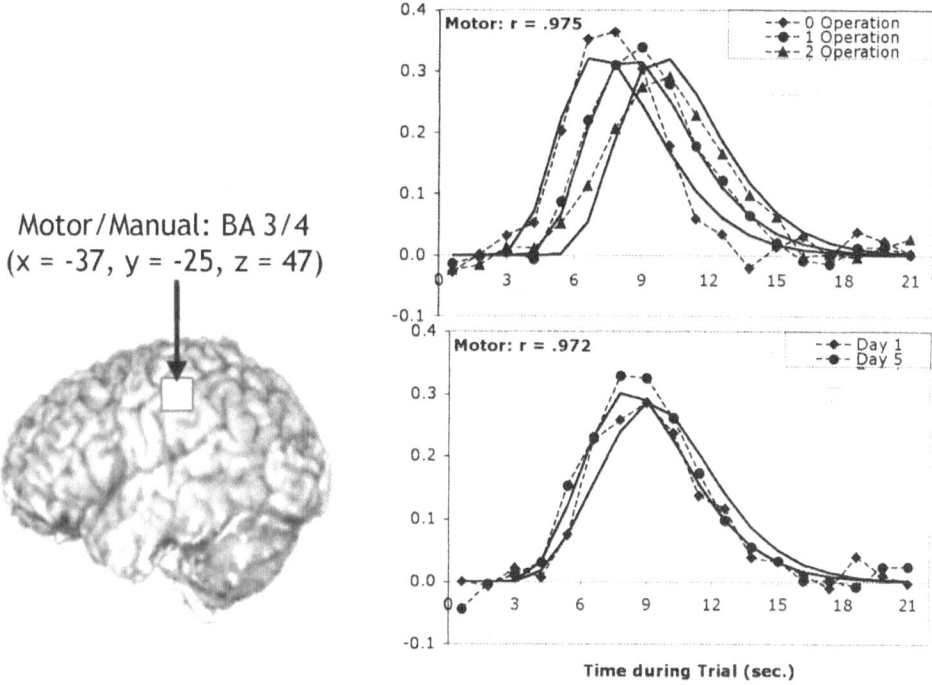

Motor/Manual: BA 3/4
(x = -37, y = -25, z = 47)

Fig. 6. Use of module behavior to predict percent increase in BOLD response in various regions: (a) Manual module predicts motor region; (b) Retrieval Module predicts prefrontal region; (c) Control/Goal module predicts anterior cingulate region; (d) Imaginal/Problem State module predicts parietal region; (e) Procedural module predicts

Table 2 gives the estimated parameters for the BOLD response and Fig. 6 shows how well this model predicts the BOLD responses in the six conditions achieved by crossing day and number of steps of transformation for each of the five associated regions. To simplify matters and to make the functions more comparable, the exponent of the BOLD response was set to 3 for all regions. To keep the data presentation readable and get better estimates, Fig. 6 either averages over days or over conditions.[5]

5.3. Characterizing the differences among the brain regions

The first impression one probably gets from Fig. 6 is that the BOLD responses for the five regions look a lot alike. All show a characteristic hemodynamic response that goes up and comes down with the trial structure. Furthermore, most regions show a stronger response for more transformations and a stronger response on Day 1. This is quite characteristic of imaging results where disparate regions of the brain give quite similar responses to the material. Without a strong theory to guide one's expectations, one is in danger of missing the differences and concluding that the whole brain (or at least those regions that respond—not all regions in the brain respond to the task structure in this experiment) is reflecting a global response to the task. However, if one knows where to look, there are characteristic differences. Although this one experiment does not reveal all the differences in the behavior of all five regions, it does reflect

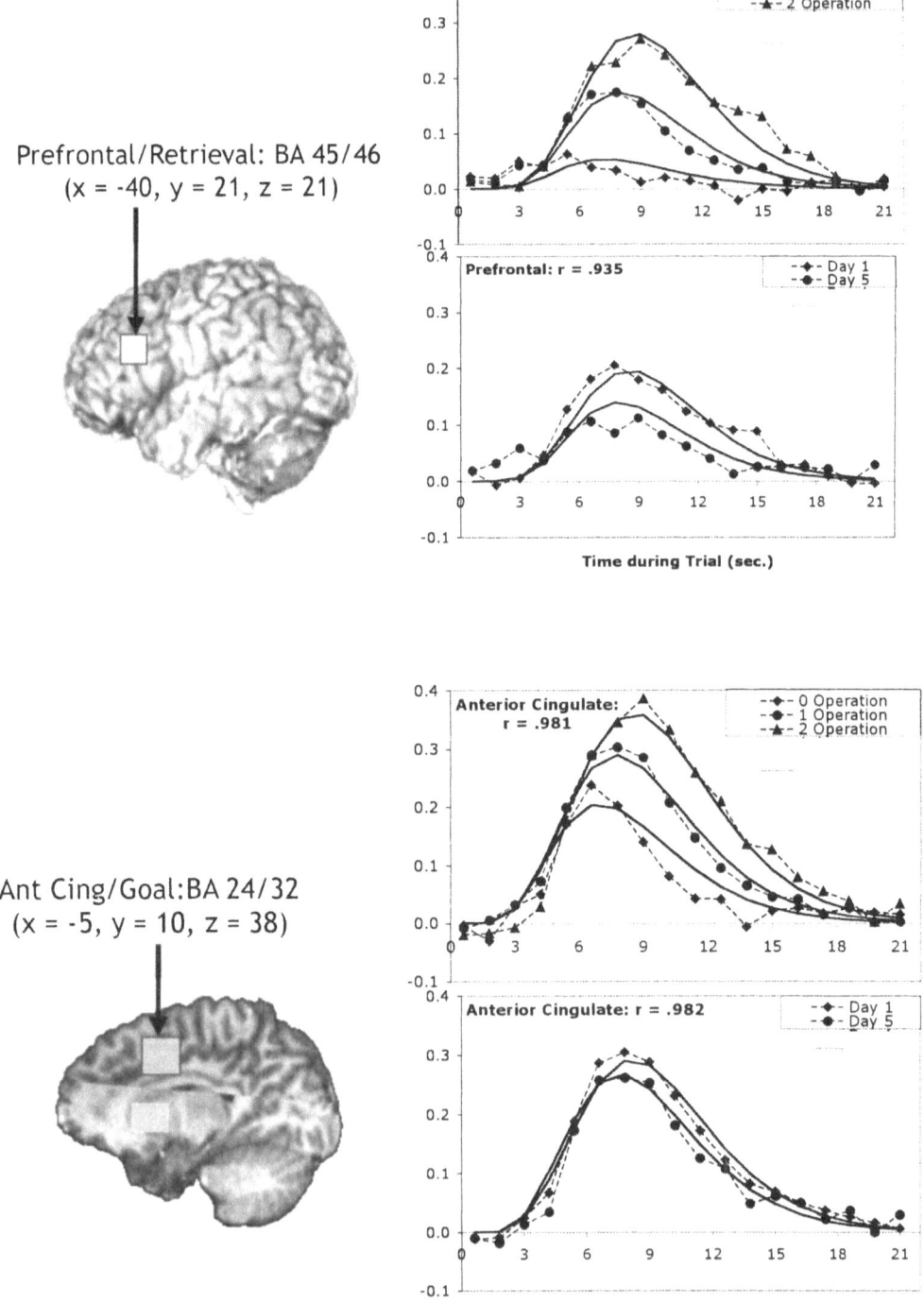

Prefrontal/Retrieval: BA 45/46
(x = -40, y = 21, z = 21)

Ant Cing/Goal:BA 24/32
(x = -5, y = 10, z = 38)

Fig. 6. *(continued)*

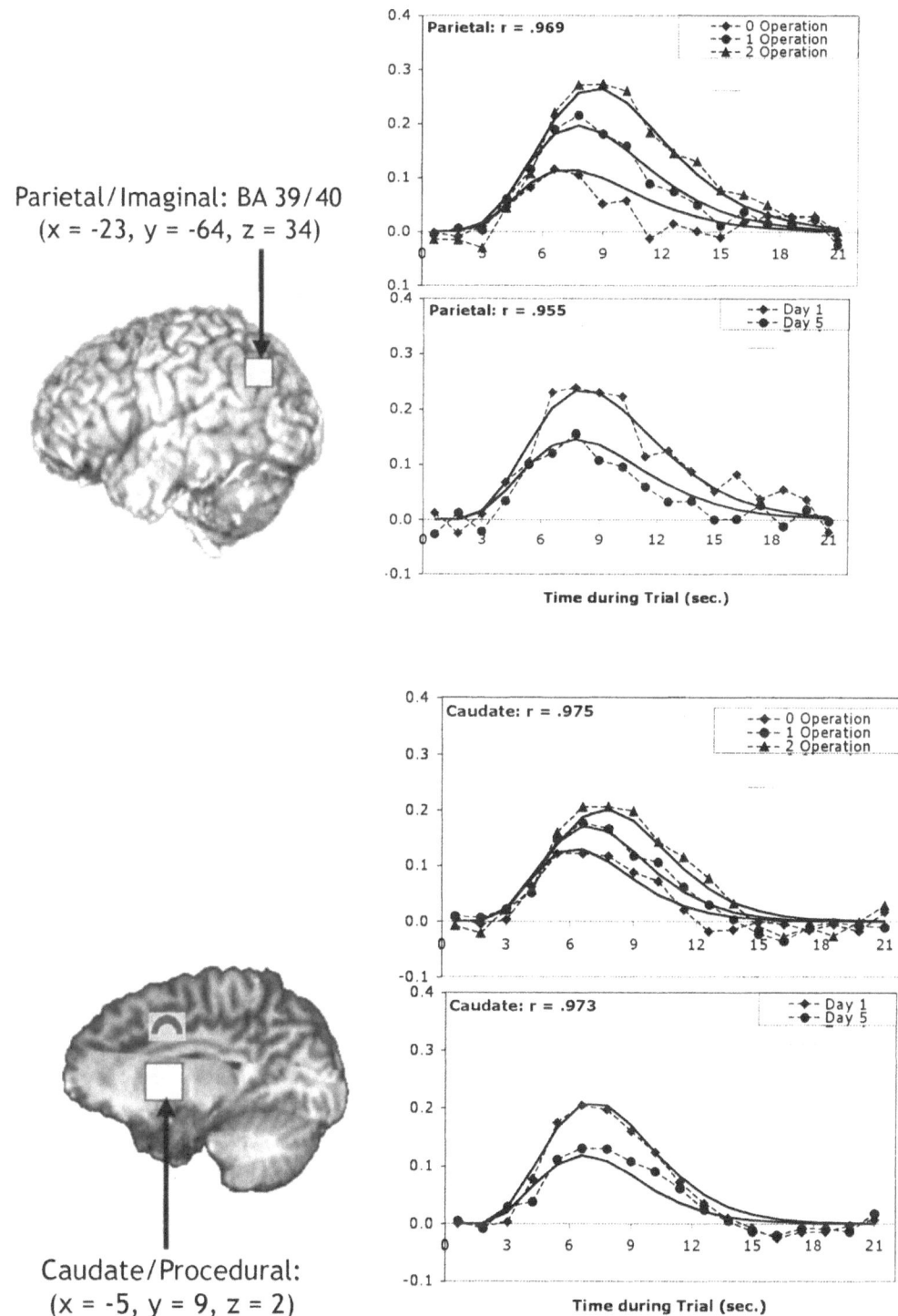

Fig. 6. *(continued)*

many of the important differences that we have identified over our experiments. These are enumerated in the following paragraphs.

First, and most apparent, in Fig. 6a the motor region is giving basically the same hemodynamic response in all conditions. The effect of the slower conditions is to delay when that hemodynamic response occurs. This is what would be expected given a relatively strong understanding of what regions of the brain control the hand. Although the motor region is transparently giving a different response than the other regions (both on theoretical and observational grounds), its correlation with the BOLD responses in other regions averages .66. Thus even it might be confused with the other regions unless one had a theory to tell one where to look to find the relevant differences.

Two of the other regions have distinct signatures. The prefrontal region (Fig. 6b) is distinguished by the very weak response it generates in the case of 0 steps. According to the model this case involves some brief retrievals of instructions but no retrieval of number facts. We have often modeled this condition by assuming no retrieval and predicted a flat function, but a slight rise can be discerned. The striking feature of the anterior cingulate (Fig. 6c) is that there is almost no effect of learning, whereas there is a robust effect of number of steps on magnitude of the response. The goal component in ACT–R is engaged in maintaining the state at points where the system is engaged in a retrieval of an arithmetic fact (this is because the retrieval buffer cannot be used to hold the next step). Every time it engages in retrieval of an arithmetic task it must note this so that it will wait for the fact before going on. Once the fact is retrieved it must reset the state so that it can proceed with unwinding. Thus, the number of retrieval operations is one factor influencing the number of state-setting operations in the goal buffer. The number of arithmetic retrievals changes in this experiment with the number of steps in solving the equation because each step requires retrieval of a fact. However, there is little reduction in these retrievals with practice. In principle, with enough practice they would eventually drop out, but there are so many individual facts that they just do not repeat enough in equation solving. In other research on learning (Qin et al., 2003) where retrievals did not involve a large database, we did find a substantial learning effect over 5 days in the anterior cingulate.

The other two regions (the parietal in Fig. 6d and the caudate in Fig. 6e) can be distinguished from the other three regions because they lack the features that identify the other three. However, there is little difference in the response that we see in these two regions. They approximately reflect the average response of all the areas, showing substantial effects of both number of steps of transformations and delay. There is a subtle difference between the two with the caudate showing a relatively larger effect of days and the parietal showing a relatively larger effect of steps. The caudate is fit according to the number of rules, which naturally increases with steps and decreases with days. These steps often are accompanied by changes in the problem representation, and this is why the two regions are so strongly correlated. We find differences between these two regions in experiments that vary modality of presentation from visual to aural with the parietal responding less to auditory presentation than visual and the caudate responding more (Sohn et al., 2005). Note in the comparisons of Fig. 6d and 6e, that the caudate gives a relatively weak response and has a poorer signal-to-noise ratio. This is unfortunate because according to the theory it should be the one region that is involved in all cognitive tasks, reflecting the number of production rules fired.

5.4. Assessing goodness of fit

The figures contain measures of correlations between the predictions and observed behavior. These are averaged over either days or operations, but Table 2 gives correlations among all 108 points for each region. Although this is a conventional measure of quality of fit, it has a number of problems. For instance, correlation is only sensitive to whether the shapes match up and not to whether the actual predicted numbers match up.

The quantitative correspondence can be assessed by the chi-square statistics in the table, which measure the degree of mismatch against the noise in the data. They are calculated as

$$\chi^2 = \frac{\sum_i (\hat{X}_i - \bar{X}_i)^2}{S_{\bar{X}}^2}$$

where the denominator is estimated from the interaction between conditions and participants. This has 105 *df*, calculated as 108 minus the three parameters estimated for the BOLD function. By this measure all of the areas are being modeled as well as can be expected because they all yield nonsignificant chi-squares (it would have to be 130 or greater to be significant at the .05 level). However, the chi-square statistic is not a perfect measure of fit for a number of reasons. First, it depends on the assumption that errors in individual points are independent, which is unlikely in this case. Second, it depends on the assumption that the gamma function is exactly the correct characterization of the BOLD response. Both of these problems reflect the fact that this measure may weight fitting the exact shape of the curves too much.

Anderson et al. (2003) offered an alternative measure of goodness of fit that avoids this concern with curve shape and parameter estimation. It simply tests whether the areas under the curves are in proportion to the time a module is active. They proposed calculating the following measure of proportionality:

$$\text{Proportionality} = \frac{\left(\sum_i T_i A_i\right)^2}{\sum_i T_i^2 \sum_i A_i^2}$$

where T_i is the amount of time a region is engaged in a condition and A_i is the amount of area under the BOLD response. In this experiment, the summations are over the six conditions defined by crossing the three levels of equation complexity with the 2 days of practice. This is like R^2 in some ways but has some differences. As a simple example of the differences consider the relation between the numbers 10, 11, and 12 and 0, 1, and 2. They have an R^2 of 1 but a proportionality of only .67 because the first set of numbers is almost equal proportionally, but the ratio of the second set is quite different. As another example, the numbers 10, 11, and 12 and the numbers 10, 12, and 10 have an R^2 of 0 but a proportionality of .987. One can calculate a degree of proportional misfit, which we call *misproportionality*, as 1 − Proportionality. These misproportionalities are reported in Table 3a, and one can observe that in all cases the region is best fit by its assigned module.

Table 3
Correlation between various modules and the BOLD response in various brain regions

	Motor	Prefrontal	Cingulate	Parietal	Caudate
a. Misproportionalities					
Manual	0.011	0.308	0.100	0.263	0.119
Retrieval	0.326	0.017	0.059	0.041	0.061
Goal	0.103	0.119	0.006	0.113	0.040
Imaginal	0.212	0.069	0.026	0.035	0.033
Procedural	0.190	0.109	0.049	0.037	0.023
b. Chi Squares					
Manual	88.93	452.05	724.66	426.40	333.89
Retrieval	493.22	82.60	350.32	101.88	133.13
Goal	255.91	194.94	123.27	171.74	111.01
Imaginal	384.66	125.66	210.47	95.21	101.82
Procedural	347.05	163.76	286.28	114.93	81.03

As a different approach, Table 3b reports the outcome of trying to fit each module to each region's activation profile and calculating a chi-square measure of misfit. With 105 *df* the 5-percentile tails for the chi-square distribution are at 82 and 130. As we noted with respect to Table 2, all the modules give acceptable fits (less than 130) to their ascribed regions. A few other modules give acceptable fits to other regions, although not as good. In particular, the modules other than the manual module all give approximately equal fits to the parietal and caudate regions. As noted, these regions approximately show the average response of all the regions. Note in Table 3 that the misproportionalities almost perfectly predict the chi-squares down a particular column (a column holds constant the noise in the data that determines the denominator for the chi-square).

In summary, the good fit of the model to the BOLD responses does not depend on the estimation of the parameters that characterize the BOLD function. Part (a) of Table 3 establishes a parameter-free measure of strength of association between ACT–R component and brain region and part (b) establishes that parameter estimation cannot make other components fit other brain regions. Except for trying to distinguish between the parietal and caudate, whose responses were not well discriminated by this experiment, the proposed associations provide a much better explanation of the data than any alternative set of associations.

6. The capacity for re-representation: A uniquely human trait?

Having now analyzed in some detail the nature of one instance of algebra symbol manipulation, I would like to close by reflecting on the question of what in these processes might be uniquely human. For this purpose it would be useful to describe an example of serial behavior that has been observed in the Rhesus Macaques monkey (Terrace, Son, & Brannon, 2003). In the experiment reported by Terrace et al., monkeys learned four 7-item lists of pictures. On any particular trial the monkeys were shown one set of seven pictures randomly arrayed on the screen, and they had to select them in the correct order. They were able to achieve over 65% correct reproduction of the entire lists—a number that would compare favorably with humans in similar circumstances. Terrace et al. offered a number of varieties of evidence to argue that the monkeys are operating off a declarative representation of the list order and not some type of procedural representation. For instance, monkeys can correctly order two items from different lists that they have never seen paired before. These serial tasks are interesting because they bear certain superficial similarities to algebra symbol manipulation. In algebra symbol manipulation a child is shown one array of symbols (the equation) and must produce another array of symbols (the solution—in our task we reduced this to a single key press, but it is typically more complex as in the example of writing of the quadratic expression at the beginning of this article). Similarly, the monkey is shown one array of symbols and must produce a sequence of symbols or actions.

Although the performance of monkeys in these serial tasks is in many ways remarkable, there is a significant difference in the behavioral capacity involved in transforming algebraic equations and that involved in manipulating serial lists. The most obvious difference is in the generativeness of the child's algebraic capacity. The child is capable of responding to an arbitrary number of new expressions. Even simply using the unwind strategy to solve equations, children are capable of solving an infinite number of equations. Moreover, the generativeness in algebraic symbol manipulation goes beyond this—for instance, a child can factor, expand, and do many other operations to transform one string of symbols into another. This kind of generativeness has much in common with the generativeness of natural language, the most frequently mentioned instance of human intellectual superiority. However, as noted in the introduction, unlike language, the formal properties of algebraic manipulation are more or less completely understood. Also as the reported experiment illustrates, it is much easier to experimentally study the learning of algebra.

Although the differences between the child's algebraic symbol manipulation and the monkey's serial reproduction may seem obvious, the challenge is to identify what in the ACT–R architecture is associated with this difference. Before addressing this question, I should acknowledge an alternative hypothesis that the difference just reflects prior knowledge: Children successful at algebra understand a number system, know their number facts, know how to parse these expressions, and know how to follow instructions. The only factor in this list that can be discounted with certainty is the importance of number knowledge because people are quite capable of learning artificial algebras (e.g., Blessing & Anderson, 1996; Qin et al., 2003) that have no numeric reference. Also, recent research has expanded upward our estimate of primates' understanding of number (see Hauser & Spelke, in press, for a review). The other differences might support the argument that monkeys would be capable of doing algebra if it were practical to teach them this skill and its prerequisite background knowledge. I cannot disprove this possibility, but it seems unlikely. Moreover, it does turn out that there is a significant ACT–R architectural capability required for algebraic manipulation that is not required for the serial reproduction tasks.

ACT–R models for the serial reproduction tasks require visual, manual, and retrieval buffers that work in formally similar ways to the models for the algebraic tasks. These, then, cannot be the source of the differences. However, these buffers in themselves do not allow the mental re-representation that is key to algebraic symbol manipulation. Although children can perform these re-representations on paper, writing out transformation after transformation, and thus saving themselves the need for mental re-representation, they prefer to do it mentally as long as things do not get so complex as to exceed their capacity to hold a representation of the critical algebraic material.

Fig. 7 illustrates the transformations in Fig. 3, somewhat simplified and designed to make salient the key architectural issues rather than to be faithful to all the details of the learning simulation. As in Fig. 3 the equation being solved is $7*x + 3 = 38$. The simplified mental image of the equation just holds the intermediate result, but it is the critical piece of information in that it is what is not supported by external information. For instance, at one point the image in Fig. 7 holds an internal representation of 35 that is intermediate between the original equation and the final answer of 5. Being able to hold onto such an internal representation, detached from either stimulus or action, is critical to the model's algebraic competence. It is tempting to point to the parietal cortex as what is enabling this algebra problem solving, because the parietal cortex is what seems to be holding the image of the intermediate result. The region of parietal cortex that

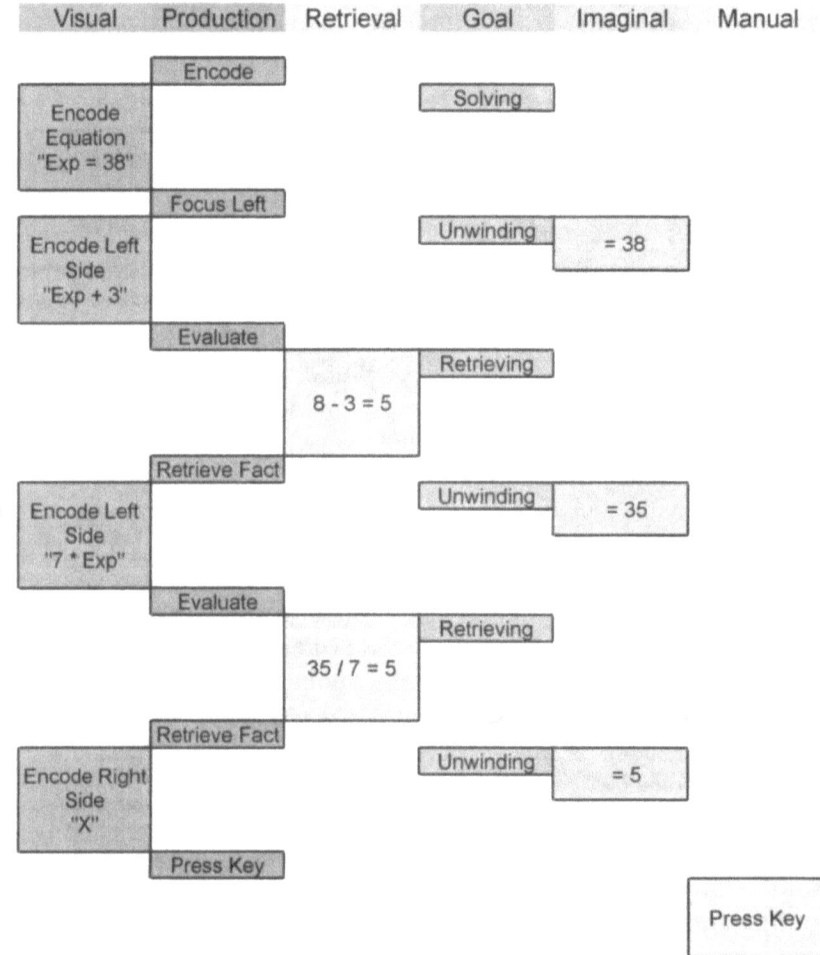

Fig. 7. A representation of the basic buffer operations required to implement the unwind strategy in ACT–R to solve the equation $7*x + 3 = 38$.

corresponds to the imaginal module does not appear to have a homologue in the monkey brain (Zilles & Palomero-Gallagher, 2001).

Although there may be some special properties to the human ability to hold such intermediate results, it is not totally discontinuous from the ability of the monkey. This becomes apparent when one tries to develop an ACT–R model for the monkey task of ordering two items from two lists. Terrace et al. (2003) showed a generative capacity to the monkey's serial knowledge in that it can take a pair of elements from different lists, which it has not seen together, and correctly order it with high accuracy. Fig. 8 is a similar flowchart for a putative ACT–R model that I developed for this ordering task. The model assumes that the monkey retrieves the location of each item in the pair and creates an image that synthesizes the two locations and then picks the item that is first in this image. Although the imaginal ability in this example may not have all of the flexibility of human imagery in equation manipulation, it seems essential to be able to have some internal synthesis of the two objects to make an appropriate decision. I could not figure a way to do this in ACT–R without such an internal representation. A comparison of Figs. 7 and 8 should make clear that the two tasks do not differ in their capacity demands on the imaginal representation. Both require a relatively small amount to be held in this working memory. In

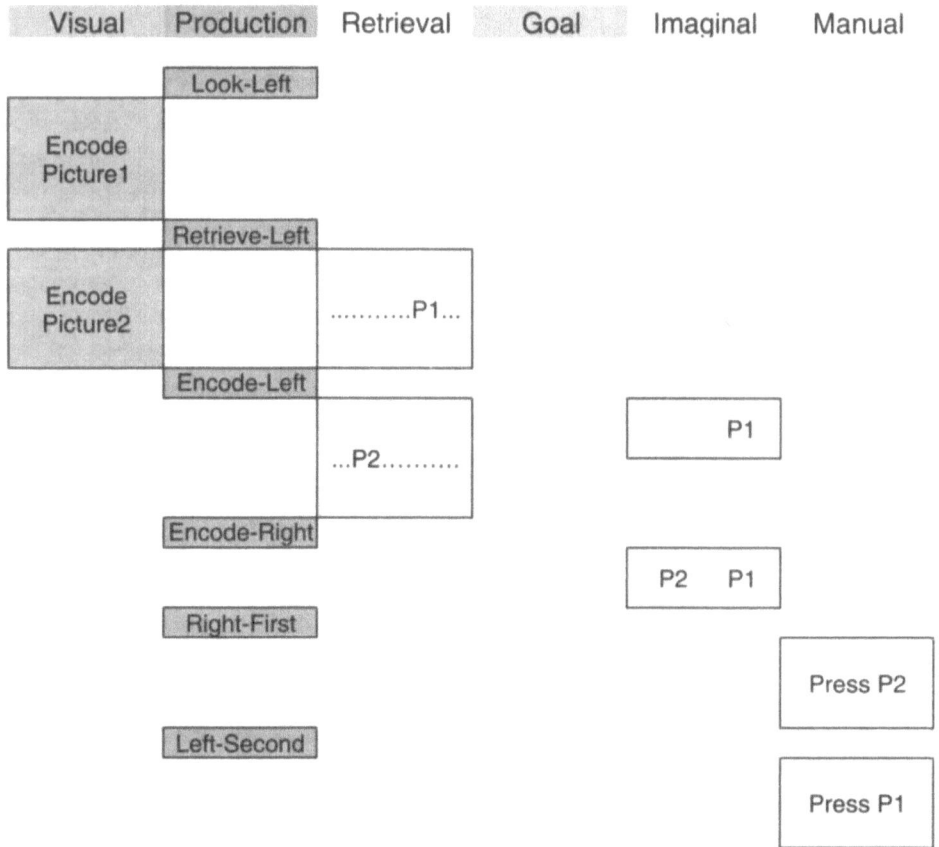

Fig. 8. A representation of the basic buffer operations in ACT–R required to implement the serial ordering task.

fact, the algebra task really requires holding just one number at any time, whereas two items have to be synthesized in the image for the serial task.

Comparing Figs. 7 and 8, however, reveals a striking difference. The model in Fig. 8 does not require any state tests against the goal buffer. I could have built a model that included such state tests (and perhaps such models are appropriate for humans doing this task), but it was unnecessary. The conditions for the firing of the individual productions are determined by the states of the other buffers. Specifically, the presentation of the stimulus is the condition for the look-left production; the encoding of the left element is the condition for the retrieve-left production that requests the position of the left element in its list; the retrieval of an element and an empty image is the condition for the encode-left production that positions the left element in the image and requests retrieval of the right element; the retrieval of an element and an incomplete image is the condition for the encode-right production that similarly places the right element; the completion of the image is the condition for the production that selects the first element; and the manual selection of that element is the condition for the second production that selects the second element.

In contrast, because of the iterative nature of the unwind algorithm, it is not possible to find unique states of the nongoal modules for each production in the algebra model. The model is faced with multiple situations where it has focused on an element in the equation, has retrieved an arithmetic fact, and has an image of an intermediate result. Without the help of the control element in the goal it would not know whether it is time to retrieve another arithmetic fact or perform another transformation of the equation. Therefore, it sometimes skips retrievals or transformations and other times repeats them. In this model repeats are innocuous, but skips mean it fails to solve the problem. For instance, if the model that does not use the control information, faced with the equation $3x + 9 = 15$, it sometimes responds 6 because it omitted retrieving $6/3 = 2$ or because it omitted to use the fact when retrieved.[6]

This goal buffer has been tentatively associated with the anterior cingulate. The anterior cingulate is particularly active in studies where participants have to direct their behavior in a way that violates typical response tendencies. As noted in the introduction, it is particularly active in dealing with conflict in the Stroop task during color naming (Carter et al., 2000). We discussed there the various theories of what is behind this activity. The anterior cingulate has undergone major evolutionary changes that are only found in humans and the closely related great apes (Allman, Hakeem, Erwin, Nimchinsky, & Hof, 2001). These changes, which include a new class of spindle-shaped cells in strongest concentration in the human anterior cingulate, appear to be related to the ability to achieve appropriate behavior in the presence of conflicting stimuli.

So where does this leave the question of what enables the human-unique aspects of algebra problem solving? The critical mental ability seems to be that of re-representation in situations where neither the external situation nor the other internal buffers indicate what to do next. Re-representation in the algebra example involves this alternation between retrieval and transformations of the internal representation, ending finally in a response. One needs some sort of state information to indicate what to do next in a series of steps of re-representation The anterior cingulate is not the only neural structure to have unique properties in the human brain, but it is one of the players in higher level human cognition, and the role it plays seems to be that of maintaining a control state in the presence of ambiguous or conflicting information.

Notes

1. Although the control states in this article will be simple they can contain information about steps, substeps, and other notes such as whether a borrow exists in a subtraction problem.
2. There are a number of comments to make on these instructions. First, these instructions already reflect a distillation of the basic axioms of algebra about doing the same operation to both sides and collapsing results. The basic axioms generate a large search space of possible transformations and students are typically given guidance as to how to select the appropriate transformations (as the students in our experiment were). Second, the instructions require further elaboration to deal with cases such as (6 − subexpression) = 3 or (6 / subexpression) = 3, where there is an asymmetric operator with the subexpression on the right. We did not present any such problems to our students. Third, they treat the special "+0" and "1*" constructions as special cases. Students were given instructions to do so.
3. Although this is a particularly dramatic example of production compilation, there are many other instances in Fig. 3 that I have not noted to avoid overly cluttering the figure.
4. Note that Fig. 3 sometimes gives different productions the same name—the goal in these naming conventions is just to try to indicate the basic function of the rule and different rules are learned with similar functions
5. In fact, none of the regions showed a significant interaction between practice and number of steps or between practice, number of steps, and scan.
6. These problems can be avoided if one puts control information into the image but then this looses the separation of control state and problem state.

Acknowledgments

This research was supported by the NSF grant ROLE: REC-0087396 and ONR Grant N00014–96–1–0491. I would like to thank Herbert Terrace for discussions relevant to preparation of the last section of this article and Jennifer Ferris for her comments on the manuscript.

References

Allman, J. M., Hakeem, A., Erwin, J. M., Nimchinsky, E., & Hof, P. (2001). Anterior cingulate cortex: The evolution of an interface between emotion and cognition. *Annals of the New York Academy of Sciences, 935,* 107–117.

Altmann, E. M., & Trafton, J. G. (2002). Memory for goals: An activation-based model. *Cognitive Science, 26,* 39–83.

Amos, A. (2000). A computational model of information processing in the frontal cortex and basal ganglia. *Journal of Cognitive Neuroscience, 12,* 505–519.

Anderson, J. R., Bothell, D., Byrne, M. D., Douglass, S., Lebiere, C., & Qin, Y. (2004). An integrated theory of mind. *Psychological Review, 111,* 1036–1060.

Anderson, J. R., & Douglass, S. (2001). Tower of Hanoi: Evidence for the cost of goal retrieval. *Journal of Experimental Psychology: Learning, Memory, & Cognition, 27,* 1331–1346.

Anderson, J. R., & Lebiere, C. (1998). *The atomic components of thought.* Mahwah, NJ: Lawrence Erlbaum Associates, Inc.

Anderson, J. R., Reder, L. M., & Lebiere, C. (1996). Working memory: Activation limitations on retrieval. *Cognitive Psychology, 30,* 221–256.

Anderson, J. R., Qin, Y., Sohn, M-H., Stenger, V. A., & Carter, C. S. (2003). An information-processing model of the BOLD response in symbol manipulation tasks. *Psychonomic Bulletin & Review, 10,* 241–261.

Anderson, J. R., Qin, Y., Stenger, V. A., & Carter, C. S. (2004). The relationship of three cortical regions to an information-processing model. *Cognitive Neuroscience, 16,* 637–653.

Anderson, J. R., Taatgen, N. A., & Byrne, M. D. (in press). Learning to achieve perfect time sharing: Architectural implications of Hazeltine, Teague, & Ivry (2002). *Journal of Experimental Psychology: Human Perception and Performance.*

Ashby, F. G., & Waldron, E. M. (2000). The neuropsychological bases of category learning. *Current Directions in Psychological Science, 9,* 10–14.

Baddeley, A. D. (1986). *Working memory.* Oxford: Oxford University Press.

Blessing, S., & Anderson, J. R. (1996). How people learn to skip steps. *Journal of Experimental Psychology: Learning, Memory and Cognition, 22,* 576–598.

Boyton, G. M., Engel, S. A., Glover, G. H., & Heeger, D. J. (1996). Linear systems analysis of functional magnetic resonance imaging in human V1. *Journal of Neuroscience, 16,* 4207–4221.

Buckner, R. L., Kelley, W. M., and Petersen, S. E. (1999). Frontal cortex contributes to human memory formation. *Nature Neuroscience, 2,* 311–314.

Cabeza, R., Dolcos, F., Graham, R., & Nyberg, L. (2002). Similarities and differences in the neural correlates of episodic memory retrieval and working memory. *Neuroimage, 16,* 317–330.

Carter, C. S., MacDonald, A. M., Botvinick, M., Ross, L. L., Stenger, V. A., Noll, D., et al. (2000). Parsing executive processes: Strategic versus evaluative functions of the anterior cingulate cortex. *Proceedings of the National Academy of Sciences of the U.S.A., 97,* 1944–1948.

Cohen, M. S. (1997). Parametric analysis of fMRI data using linear systems methods. *NeuroImage, 6,* 93–103.

Dale, A. M., & Buckner, R. L. (1997). Selective averaging of rapidly presented individual trials using fMRI. *Human Brain Mapping, 5,* 329–340.

Dehaene, S., Piazza, M., Pinel, P., & Cohen, L., (2003). Three parietal circuits for number processing. *Cognitive Neuropsychology, 20,* 487–506.

Dehaene, S., Posner, M. I., & Tucker, D. M. (1994). Localization of a neural system for error detection and compensation. *Psychological Science, 5,* 303–305.

D'Esposito, M., Piazza, M., Detre, J. A., Alsop, D. C., Shin, R. K., Atlas, S., et al. (1995). The neural basis of the central executive of working memory. *Nature, 378,* 279–281.

Donaldson, D. I., Petersen, S. E., Ollinger, J. M., & Buckner, R. L. (2001). Dissociating state and item components of recognition memory using fMRI. *NeuroImage, 13,* 129–142.

Fletcher, P. C., & Henson, R. N. A. (2001). Frontal lobes and human memory: Insights from functional neuroimaging. *Brain, 124,* 849–881.

Falkenstein, M., Hohnbein, J., & Hoorman, J. (1995). Event related potential correlates of errors in reaction tasks. In G. Karmos, M. Molnar, V. Csepe, I. Czigler, & J. E. Desmedt (Eds.), *Perspectives of event-related potentials research* (pp. 287–296). Amsterdam: Elsevier Science.

Frank, M. J., Loughry, B., & O'Reilly, R. C. (2000). *Interactions between frontal cortex and basal ganglia in working memory: A computational model* (Tech. Rep. 00-01). Boulder, CO: University Colorado, Institute of Cognitive Science.

Glover, G. H. (1999). Deconvolution of impulse response in event-related BOLD fMRI. *NeuroImage, 9,* 416–429.

Hauser, M. D., & Spelke, E. S. (in press). Evolutionary and developmental foundations of human knowledge: A case study of mathematics. In M. Gazzaniga (Ed.), *The cognitive neurosciences III.* Cambridge, MA: MIT Press.

Hikosaka, O., Nakahara, H., Rand, M. K., Sakai, K., Lu, Z., Nakamura, K., et al. (1999). Parallel neural networks for learning sequential procedures. *Trends in Neuroscience, 22,* 464–471.

Houk, J. C., & Wise, S. P. (1995). Distributed modular architectures linking basal ganglia, cerebellum, and cerebral cortex: Their role in planning and controlling action. *Cerebral Cortex, 2,* 95–110.

Koedinger, K. R., Anderson, J. R., Hadley, W. H., & Mark, M. (1997). Intelligent tutoring goes to school in the big city. *International Journal of Artificial Intelligence in Education, 8,* 30–43.

Lebiere, C., & Anderson, J. R. (1993). A connectionist implementation of the ACT–R production system. In *Proceedings of the Fifteenth Annual Conference of the Cognitive Science Society* (pp. 635–640). Mahwah, NJ: Lawrence Erlbaum Associates, Inc.

Lepage, M., Ghaffar, O., Nyberg, L., & Tulving, E. (2000). Prefrontal cortex and episodic memory retrieval mode. *Proceedings of National Academy of Sciences U.S.A., 97,* 506–511.

MacDonald, A. W., Cohen, J. D., Stenger, V. A., & Carter, C. S. (2000). Dissociating the role of dorsolateral prefrontal and anterior cingulate cortex in cognitive control. *Science, 288,* 1835–1838.

Meyer, D. E., & Kieras, D. E. (1997). A computational theory of executive cognitive processes and multiple-task performance. Part 1. Basic mechanisms. *Psychological Review, 104,* 2–65.

Norman, D. A., & Shallice, T. (1986). Attention to action: Willed and automatic control of behavior. In R. J. Davidson, G. E. Schwartz, & D. Shapiro (Eds.), *Consciousness and self-regulation: Vol. 4. Advances in research and theory* (pp. 1–18). New York: Plenum.

Pashler (1994). Dual-task interference in simple tasks: Data and theory. *Psychological Bulletin, 116,* 220–244.

Paus, T., Koski, L., Caramanos, Z., & Westbury, C. (1998). Regional differences in the effects of task difficulty and motor output on blood flow response in the human anterior cingulate cortex: A review of 107 PET activation studies. *Neuroreport, 9,* R37–R47.

Poldrack, R. A., Prabakharan, V., Seger, C., &Gabrieli, J. D. E. (1999). Striatal activation during cognitive skill learning. *Neuropsychology, 13,* 564–574.

Posner, M. I., & Dehaene, S. (1994). Attentional networks. *Trends in Neurosciences, 17,* 75–79.

Posner, M. I., & DiGirolamo, G. J. (1998). Executive attention: Conflict, target detection and cognitive control. In R. Parasuraman (Ed.), *The attentive brain* (pp. 401–423). Cambridge, MA: MIT Press.

Qin, Y., Sohn, M-H., Anderson, J. R., Stenger, V. A., Fissell, K., Goode, A., et al. (2003). Predicting the practice effects on the blood oxygenation level-dependent (BOLD) function of fMRI in a symbolic manipulation task. *Proceedings of the National Academy of Sciences of the U.S.A., 100,* 4951–4956.

Qin, Y., Anderson, J. R., Silk, E., Stenger, V. A., & Carter, C. S. (2004). The change of the brain activation patterns along with the children's practice in algebra equation solving. *Proceedings of the National Academy of Sciences of the U.S.A., 101,* 5686–5691.

Reichle, E. D., Carpenter, P. A., & Just, M. A. (2000). The neural basis of strategy and skill in sentence-picture verification. *Cognitive Psychology, 40,* 261–295.

Saint-Cyr, J. A., Taylor, A. E., & Lang, A. E. (1988). Procedural learning and neostriatal dysfunction in man. *Brain, 111,* 941–959.

Sohn, M-H., Albert, M. V., Jung, K., Carter, C. S., & Anderson, J. R. (2004). *Pay now or pay later: Preparatory conflict resolution in the anterior cingulate cortex and the prefrontal cortex.* Poster session presented at the 11th Annual Meeting of the Cognitive Neuroscience Society, San Francisco.

Sohn, M-H., Goode, A., Stenger, V. A, Carter, C. S., & Anderson, J. R. (2003). Competition and representation during memory retrieval: Roles of the prefrontal cortex and the posterior parietal cortex. *Proceedings of National Academy of Sciences, 100,* 7412–7417.

Sohn, M-H., Goode, A., Stenger, V. A, Jung, K-J., Carter, C. S., & Anderson, J. R. (2005). An information-processing model of three cortical regions: Evidence in episodic memory retrieval. *NeuroImage, 25,* 21–33.

Taatgen, N. A. (2005). Modeling parallelization and speed improvement in the skill acquisition: From dual tasks to complex dynamic skills. *Cognitive Science, 29,* 421–455.

Taatgen, N. A., & Anderson, J. R. (2002). Why do children learn to say "Broke"? A model of learning the past tense without feedback. *Cognition, 86,* 123–155.

Terrace, H. S., Son, L. K., & Brannon, E. M. (2003). Serial expertise of rhesus macaques. *Psychological Science, 14,* 66–73.

Wagner, A. D., Maril, A., Bjork, R. A., & Schacter, D. L. (2001). Prefrontal contributions to executive control: fMRI evidence for functional distinctions within lateral prefrontal cortex. *NeuroImage, 14,* 1337–1347.

Wagner, A. D., Paré-Blagoev, E. J., Clark, J., & Poldrack, R. A. (2001). Recovering meaning: Left prefrontal cortex guides controlled semantic retrieval. *Neuron, 31,* 329–338.

Wise, S. P., Murray, E. A., & Gerfen, C. R. (1996). The frontal cortex-basal ganglia system in primates. *Critical Reviews in Neurobiology, 10,* 317–356.

Zilles, K., & Palomero-Gallagher, N. (2001). Cyto-, myelo-, and receptor architectonics of the human parietal cortex. *Neuroimage, 14,* 8–20.

Cognitive Science 29 (2005) 343–373

Rational Analyses of Information Foraging on the Web

Peter Pirolli

PARC, Information Sciences and Technologies Laboratory

Received 28 April 2004; received in revised form 13 December 2004; accepted 5 January 2005

Abstract

This article describes rational analyses and cognitive models of Web users developed within information foraging theory. This is done by following the rational analysis methodology of (a) characterizing the problems posed by the environment, (b) developing rational analyses of behavioral solutions to those problems, and (c) developing cognitive models that approach the realization of those solutions. Navigation choice is modeled as a random utility model that uses spreading activation mechanisms that link proximal cues (*information scent*) that occur in Web browsers to internal user goals. Web-site leaving is modeled as an ongoing assessment by the Web user of the expected benefits of continuing at a Web site as opposed to going elsewhere. These cost–benefit assessments are also based on spreading activation models of information scent. Evaluations include a computational model of Web user behavior called Scent-Based Navigation and Information Foraging in the ACT Architecture, and the Law of Surfing, which characterizes the empirical distribution of the length of paths of visitors at a Web site.

Keywords: Information foraging, Information scent, World Wide Web, Rational analysis, ACT–R, SNIF–ACT

1. Introduction

A prevalent strategy for engaging and adapting to modern physical and social environments is to acquire and use relevant external information. People (and technology designers) are faced with developing adaptive ways of acquiring and using external content to gain knowledge that improves decision making and problem solving. *Information foraging theory* (Pirolli & Card, 1999) addresses human–information interaction (HII) involving modern technologies such as the World Wide Web. The theory is concerned with human behavior and technology involved in gathering information for some purpose, such as making a medical decision, finding a restaurant, or solving a programming problem. This article presents an application of information foraging theory to the behavior observed in finding information on the Web. The focus

---

Requests for reprints should be sent to Peter Pirolli, PARC, Information Sciences and Technologies Laboratory, 3333 Coyote Hill Road, Palo Alto, CA 94304. E-mail: pirolli@parc.com

is on understanding the behavior of users who are not expert in finding domain-specific information (cf. Bhavnani, 2002).

The human propensity to gather and use information to adapt to everyday problems in the world is a core piece of human psychology that has been largely ignored in cognitive studies. G. A. Miller (1983) argued that the human species might fruitfully be viewed as a kind of *informavore:* a species that hungers for information in order to gather it and store it as a means for adapting to the world. Humans, of course, are extreme in their reliance on information, with language and culture, and now modern technology, providing media for transmission within and across generations. Humans are the *Informavores Rex* of this era.

To develop theory and models of human information gathering behavior, information foraging theory has adopted the *rational analysis* program initiated by Anderson (1989, 1990, 1991). The rational analysis approach involves a kind of reverse engineering in which the theorist asks (a) *what* environmental problem is solved, (b) *why* is a given behavioral strategy a good solution to the problem, and (c) *how* is that solution realized by cognitive mechanism? The products of this approach include (a) characterizations of the relevant goals and environment, (b) mathematical rational choice models (e.g., optimization models) of idealized behavioral strategies for achieving those goals in that environment, and (c) computational cognitive models. This methodology is founded on the heuristic assumption that evolving, behaving systems are well designed (rational) for fulfilling certain functions in certain environments. Rational analysis is a variant form of an approach called *methodological adaptationism* that has also shaped research programs in behavioral ecology (e.g., Mayr, 1983; Stephens & Krebs, 1986; Tinbergen, 1963), anthropology (e.g., Winterhalder & Smith, 1992), and neuroscience (e.g., Glimcher, 2003).

In keeping with the rational analysis approach, this article begins with some general empirical characterizations about the structure of the Web as an environment for information gathering behavior. Then, rational analyses are developed to characterize optimal behavioral strategies for information foraging on the Web. Surprisingly, a branch of behavioral ecology called optimal foraging theory (Stephens & Krebs, 1986), which predicts the strategies used by animals and hunter-gatherers to forage for food, has been a valuable resource (among others) for developing rational analyses of how people forage for information. Finally, this article presents empirical evaluations of the optimization models and a computational cognitive model derived from the optimization analyses called Scent-Based Navigation and Information Foraging in the ACT Architecture (SNIF–ACT).

2. Aspects of the Web environment

In this section, I review empirical observations about the Web to establish some general characterizations about task environments that involve the Web. These characterizations shape the rational analyses performed in the next section. As discussed in this section, people use the Web to acquire knowledge to improve ill-structured decision-making and problem solving. Interacting with the Web also involves costs—especially the opportunity cost of the time involved. Information foraging behavior is rational to the extent that it maximizes the value of the knowledge gained from the Web relative to the cost of interaction. Within this general framing of the value and costs of information foraging on the Web, there are two characteristic problems posed by the Web environment (and there are surely others): (a) the problem of choosing which links to follow based on available cues and (b) when to give up on a current Web locality (e.g., a Web site) and go to another.

2.1. Ill-defined problems and the value of Web content

When people are asked to report tasks they deem "significant" (Morrison, Pirolli, & Card, 2001), the majority of responses identify *ill-structured* problems (Reitman, 1964; Simon, 1973) requiring the acquisition of additional knowledge from external content sources. Ill-structured problems, such as choosing a medical treatment or buying a house typically require additional knowledge from external sources to better understand the starting state, to better define a goal, or to specify the actions that are afforded at any given state (Simon, 1973). People typically need to perform *knowledge search* (Newell, 1990) to solve their ill-structured problems (e.g., to define aspects of a problem space that permit effective or efficient problem space search). The Web is a potential source of valuable knowledge that can improve our range of adaptation because we can solve more problems, or solve problems using better approaches.

2.2. Information scent cues for Web navigation

The structure of the Web has evolved to exhibit regularities in the distribution of information resources and the navigation paths that lead to those resources. One regularity is the availability of labeled navigation links from one Web page to another (e.g., Fig. 1), and users appear to

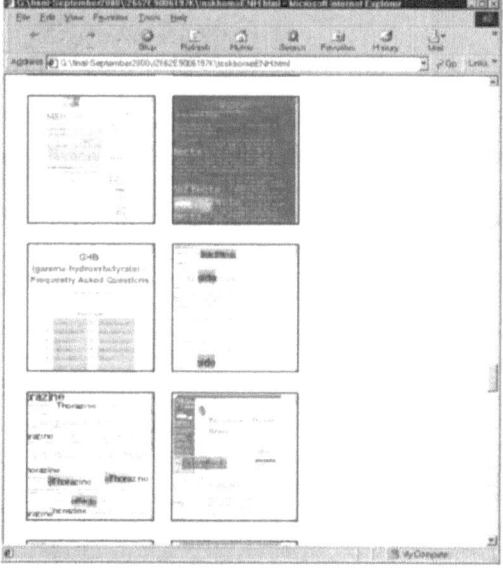

(a)   (b)

Fig. 1. Examples of information scent cues: (a) a typical set of search results in the form of textual summaries of linked documents and (b) Relevance Enhanced Thumbnails for the same set of search results.

prefer following links over other means of Web navigation (Katz & Byrne, 2003). Web page designs have evolved to associate (by human design or automated information systems) small snippets of text and graphics with such links. Those text and graphics cues are intended to represent tersely the content that will be encountered by choosing a particular link on one page and navigating to the linked page. When browsing the Web by following links, users must use these cues presented proximally on the Web pages they are currently viewing to make navigation decisions. These link cues are called *information scent.* For the Web user, there is uncertainty about the relation of the proximal cues to the linked information resources.

Fig. 1 presents some examples of information scent cues. Fig. 1a is a typical page generated by a Web search engine in response to a user query. The page lists Web pages (search results) that are predicted to be relevant to the query. Each search result is represented by its title (in blue), phrases from the hit containing words from the query, and a URL (Uniform Resource Locator). Fig. 1b illustrates an alternative form of search result representation (for exactly the same items in Fig. 1a) that is provided by *relevance-enhanced thumbnails* (Woodruff, Rosenholtz, Morrison, Faulring, & Pirolli, 2002), which combine thumbnail images of search results with highlighted text relevant to the user's query. In the complex network organization of the Web, small perturbations in the accuracy of information scent can cause qualitative shifts in the cost of browsing. This will be illustrated by application of an analysis of phase transitions in heuristic search developed by Hogg and Huberman (1987).

For a specific arrangement of Web content, interpage links, and information scent on a Web site, it is becoming possible to make model-based predictions about likely time costs of navigation—for each specific case. Examples of such models include SNIF–ACT (discussed later), Bloodhound (Chi et al., 2003), Cognitive Walkthrough for the Web (Blackmon, Polson, Kitajima, & Lewis, 2002), and MESA (C. S. Miller & Remington, 2004). The approach of Hogg and Huberman (1987), however, provides a way of characterizing the general relation between information scent and navigation costs for the asymptotic case. Consider an idealized case of browsing for information by surfing along links in the Web. The information structure generally will be a lattice of interlinked Web pages. At each visited page, the browsing will involve choosing a link to pursue from a set of presented links. Assume that the browsing takes the form of a hierarchically arranged search tree in which branches are explored, and if unproductive, the user returns to previously visited pages. To an approximation this matches observation (Card et al., 2001). Although the Web is a lattice, the search process over that lattice tends to follow a treelike form (i.e., a spanning tree is generated by the search process over a more general graph structure of the Web). This somewhat idealized case of hierarchically organized Web search is summarized in Fig. 2. The search tree may be characterized by a branching factor b corresponding to the average number of alternatives available at each decision point. The desired target information may occur at various depths, d, in the search tree (Fig. 2).

An exhaustive tree-search process would visit every leaf node. Such a full (exhaustive) search is indicated by the complete set of lines tracing the tree structure in Fig. 2. This exhaustive tree search process would visit b^d nodes. A random tree search that terminated on encountering a target (goal) node would search about half of the tree on average. Such a random search, indicated by the light lines in Fig. 2, would visit about $b^d/2$ nodes. Information

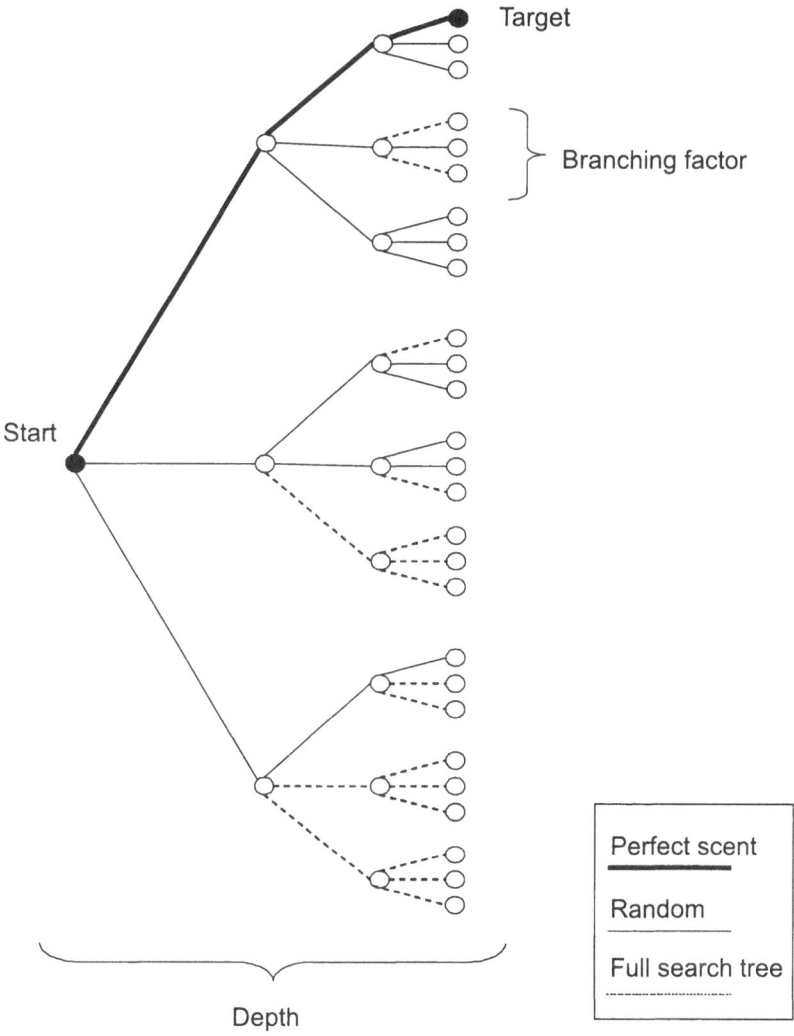

Fig. 2. Idealized search trees for heuristic search under perfect information scent (heavy solid lines) and no information scent (random, light solid lines). Dotted lines illustrate additional paths a full (exhaustive) search tree.

scent could be used as a heuristic to improve this search even further. At each node in the tree (corresponding to a Web page) some paths could be eliminated by consideration of the information scent associated with links to those paths. Improved information scent would improve the elimination of unproductive paths. If information scent were perfect, then the user would make no incorrect choices. Let us associate a false alarm factor, f, with the probability of failing to eliminate an incorrect link. At the extremes, perfect information scent would correspond to $f = 0$ (all wrong paths eliminated), and random guessing would correspond to $f = 1$ (no wrong paths eliminated). In Fig. 2 the heavily weighted line illustrates a search involving perfect information scent ($f = 0$) and the light line illustrates a random search ($f = 1$).

According to Hogg and Huberman (1987), the average number of nodes examined in such a hierarchical search process will be

$$N(d,b,f) = d + \left[\frac{(b-1)f}{2}\right]\sum_{s=1}^{d-1}\frac{1-(bf)^s}{1-bf}$$ (1)

Equation 1 captures a three-way interaction of information scent, f, branching factor, b, and depth, d. With perfect information scent ($f = 0$), the search cost is just d. In a random search ($f = 1$) the search process must visit about half the nodes on average before target information is found.

Fig. 3 shows the effects of perturbations in false alarm factor more concretely by displaying search cost functions for a hypothetical Web site with branching factor $b = 10$. Search cost refers to the number of pages a user must visit before arriving at the desired page. The curves represent cost functions for links with false alarm rates of $f = .015, .100, .125,$ and $.150.$, which is about the range observed empirically in a study of information scent cues (Woodruff et al., 2002). One can see that the search cost regime changes very little as f ranges from .015 to .100,

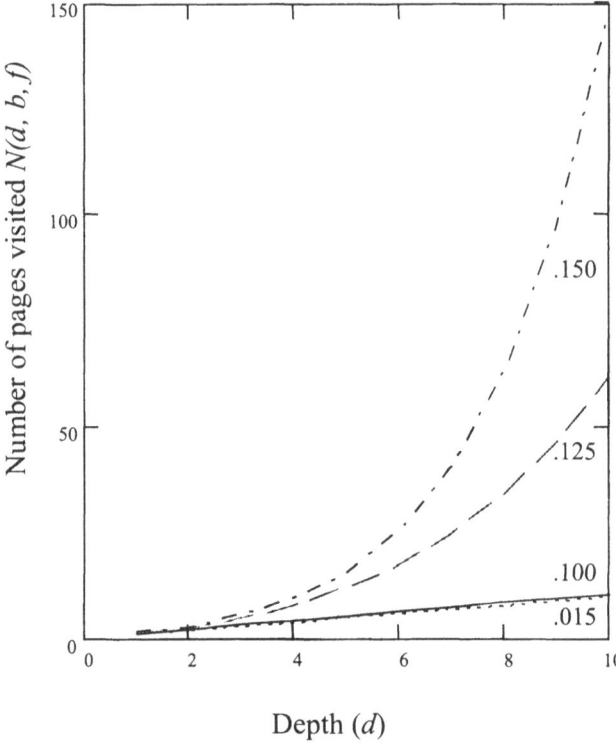

Depth (d)

Fig. 3. Effects of perturbations of false alarm rates (f; indicated by labels next to each curve) on a hypothetical Web site with branching factor $b = 10$. A qualitative shift from linear search costs to exponentially increasing costs occurs at $f = .100$.

but changes dramatically as f becomes greater than .100. Indeed, for a branching factor of $b =$ 10, there is a phase change from a linear search cost to an exponential search cost at the critical value of $f = .100$ (Hogg & Huberman, 1987). Small improvements in the false alarm factor associated with individual links can have dramatic qualitative effects on surfing large hypertext collections. Rational Web users should be motivated to maximize the accuracy of their judgments based on information scent to reduce search costs.

2.3. The hierarchical patch structure of the Web

There are a number of regularities about the distribution of links and content on the Web (e.g., Baldi, Frasconi, & Smyth, 2003). As discussed in this section, for an arbitrary task performed by a user, it appears that Web content tends to exhibit a patchy structure, in which clusters of relevant Web pages will be localized (i.e., it will be easy for a user to go from one page to another), but going from one cluster to another may require some effort. The patchy structure of the Web environment presents a kind of time allocation problem for the user: how long to forage in a patch of information before moving on to another.

One of the conventional models in the optimal foraging theory (Stephens & Krebs, 1986) deals with cases in which an organism faces an environment that has a patchy arrangement of food sources, such as birds that feed on berries that cluster on bushes that are more or less randomly scattered in the environment. The rational choice models in this area concern decisions of how long to forage in a patch (e.g., a berry bush) that exhibits diminishing returns before moving on to search for another patch. In keeping with the literature on foraging theory, I refer to groupings of information generically as *information patches*—the idea being that it is easier to navigate and process information that resides within the same patch than to navigate and process information across patches. The term *patch* suggests a locality in which within-patch navigation distances are smaller than between-patch distances. Pirolli and Card (1999) presented the classical patch model from optimal foraging theory and showed its application to user behavior in browsing a document clustering system called Scatter/Gather (Cutting, Karger, & Pedersen, 1993; Cutting, Karger, Pedersen, & Tukey, 1992).

Simon's (1962) work on the architecture of complexity argued that information systems tend to evolve toward hierarchical organizations. In part this has to do with robustness, but it also has to do with efficiency. Efficient hierarchically arranged information systems can emerge from decentralized social processes. Eiron and McCurley (2003) suggested that the link structure of the Web will tend to form localized information patches that reflect the social organizations surrounding the authorship process. For instance, the organization of groups within departments, within schools, within universities, might be reflected in a Web-site directory structure. Authors will tend to link to pages written by authors they know, and the likelihood of interauthor familiarity will decrease with tree distance in the hierarchy of the social organization, which is often reflected in the Web-site directory structure.

Empirical studies of the link structure of the Web indeed reveal a hierarchical patchy structure. In an analysis 616 million pages from 12.5 million Web sites sampled from a crawl of the Web conducted in 2002, Eiron and McCurley (2003) found that 75.0% of the links from Web pages go to other Web pages within the same Web site, and 54.8% of the within-site links go to Web pages in the same directory. Eiron and McCurley also found that the probability of occurrence of a hyperlink between two Web pages decreased exponentially with their distance in the directory structure of a Web-site host.

Web users often surf the Web seeking content related to some topic of interest, and the Web tends to be organized into topical localities. Davison (2000) analyzed the topical locality of the Web using 200,998 Web pages sampled from approximately 3 million pages crawled in 1999. Davison assessed the topical similarity of pairs of Web pages from this sample that were linked (had a link between them), siblings (linked from the same parent page), and random (selected at random). The similarities were computed by a normalized correlation or cosine measure, r, on the vectors of the word frequencies in a pair of documents (Manning & Schuetze, 1999).[1] The linked pages showed greater textual similarity ($r = .23$) than sibling pages ($r = .20$), but both were substantially more similar than random pairs of pages ($r = .02$).

Fig. 4 illustrates these general findings of Davison (2000) in a concrete case. Fig. 4 shows how topical similarity between pages diminishes with the link distance between them. To produce Fig. 4, I used data collected from the Xerox.com Web site in May, 1998, and I computed the page-to-page content similarities for all pairs of pages at minimum distances of 1, 2, 3, 4, and 5° of separation. The similarities were computed by comparing normalized correlations of vectors of the word frequencies in a pair of documents (Manning & Schuetze, 1999). Fig. 4

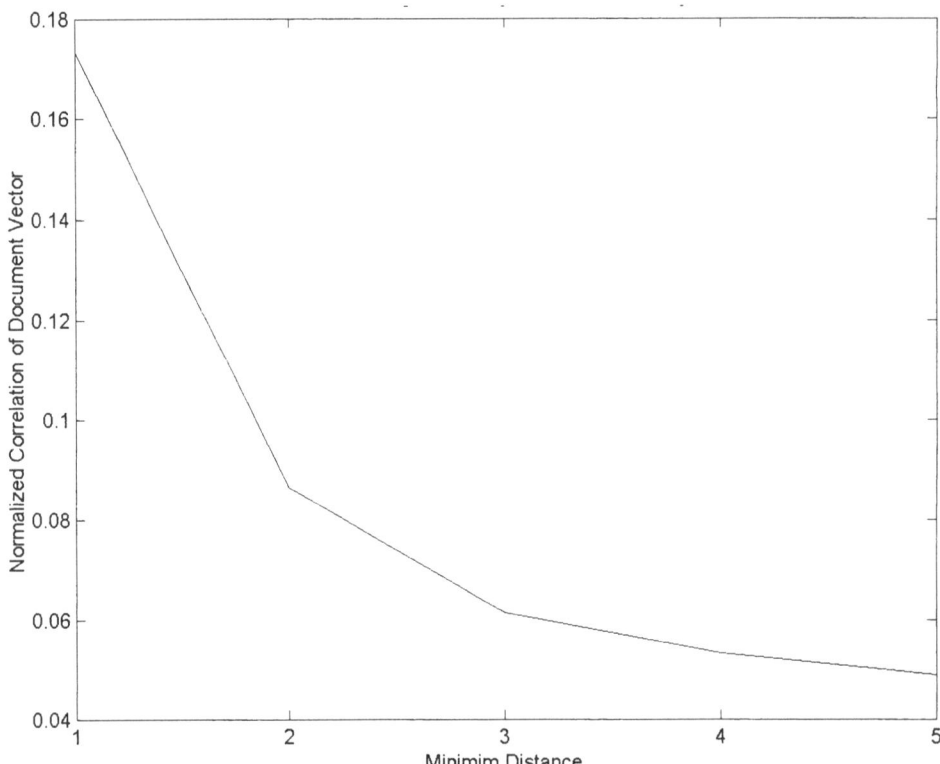

Fig. 4. The similarity (normalized correlation of document word frequency vectors) of pairs of Web pages as a function of the minimum link distances separating the pairs. Data collected from the Xerox.com Web site, May 1998.

shows that the similarity of the content of pages diminishes rapidly as a function of shortest link distance separating them.

Regardless of the ultimate evolutionary causes, it is clear that the Web is organized into a hierarchy of patches (Eiron & McCurley, 2003) and Web content is arranged into topically related patches (Davison, 2000). We may expect a rational information forager on the Web to be concerned with determining whether they are in a patch of information that is likely to yield results, or to give up and find another promising information patch.

2.4. Summary

We might assume that the adaptive forager chooses interaction methods and actions that tend to optimize the utility of information gained as a function of interaction cost, when feasible. Web users cannot have perfect knowledge of where to find desirable information. When navigating the Web by following links, users must use local proximal cues, called information scent, to make navigation choices. In general, it is expected that cues have only a probabilistic relation to distal desired information sources. The accuracy of navigation choices based on the assessments of information scent can have dramatic effects on the costs of information foraging, and an adaptive information forager might be expected to optimize such judgments and choices.

Information on the Web is arranged into hierarchical patches. Users need to detect whether a patch of information is relevant to their needs, and when to give up on a patch and seek another one. Assessing the local information scent cues over a series of visited pages might be one method for making these patch-leaving decisions. Information scent may also be a means of detecting that one is moving out of a patch of topical relevance (i.e., by detecting that the quality of information scent is dwindling as in Fig. 4).

3. Rational analyses of foraging on the Web

The term *rational analysis* was inspired by rational choice theory in economics, in which people are assumed to be rational decision makers who optimize their behavioral choices in order to maximize their goals (utility). In rational analysis, however, it is not the person who is the agent of rational choice, but rather it is the selective forces of the environment that choose better biological and behavioral designs. Anderson used rational analysis to study the human cognitive architecture by assuming that natural information processing mechanisms involved in such functions as memory (Anderson & Milson, 1989) and categorization (Anderson, 1991) were well designed by evolutionary forces to meet the problems posed by the environment (see also, Oaksford & Chater, 1998).

This section presents rational analyses of some key aspects of information foraging on the Web. The first set of rational analyses concerns the use of information scent to navigate the link structure of the Web. The second set of rational analyses concerns decisions of when to leave an information patch for another. Before presenting these two sets of rational analyses, it is worth providing a general characterization of the optimization problem facing the information forager.

3.1. The principle of the extremization of information utility as a function of interaction cost

The information forager may be viewed as operating in a *task environment* and an *information environment*. The task environment (Newell & Simon, 1972) is the scientist's analysis of those aspects of the physical, social, virtual, and cognitive environments that drive human behavior. The information environment is a tributary of external content that yields knowledge that permits people to more adaptively engage the task environment. Our particular analytic viewpoint on the information environment will be determined by the information needs that arise from the embedding task environment.

In modern society, people interact with information through technology that more or less helps them find and use the right knowledge at the right time. Information foraging theory assumes that people shape themselves and their technologies to be more adaptive in reaction to their goals and the structure and constraints of their information environments. A useful way of thinking about such adaptation is to say that

> Human-information interaction systems will tend to maximize the value of external knowledge gained relative to the cost of interaction.

Schematically, we may characterize this maximization tendency as

$$\max \left[\frac{\text{Expected value of knowledge gained}}{\text{Cost of interaction}} \right] \qquad (2)$$

This hypothesis is consistent with Resnikoff's (1989, p. 97) conclusion that natural and artificial information systems evolve toward stable states that maximize gains of valuable information per unit cost (when feasible). Cognitive systems engaged in information foraging will exhibit such adaptive tendencies, and they will prefer technologies that tend to maximize the value (or utility) of knowledge gained per unit cost of interaction. It is beyond the scope of this article to present theories of the utility of information, but discussion of this topic may be found in microeconomics (e.g., Stigler, 1961), artificial intelligence (e.g., Pearl, 1988, see especially pp. 313–327), and the foundations of cognitive science (Newell, 1990, Section 2.7).

3.2. Rational analysis of the use of information scent to navigate Web links

The rational analysis of the use of information scent assumes that the goal of the information forager is to use proximal external information scent cues (e.g., a Web link) to predict the utility of distal sources of content (i.e., the Web page associated with a Web link), and to choose to navigate the links having the maximum expected utility. This rational analysis decomposes into three parts: (a) a Bayesian analysis of the expected relevance of a distal source of content conditional on the available information scent cues, (b) a mapping of this Bayesian model of information scent onto a mathematical formulation of spreading activation, and (c) a model of rational choice that uses spreading activation to evaluate the utility of alternative choices of Web links. This rational analysis yields a spreading activation theory of utility and choice.

3.2.1. Bayesian analysis of information scent

Anderson and Milson (1989) proposed that human memory is designed to solve the problem of predicting what past experiences will be relevant in ongoing current proximal contexts, and allocating resources for the storage and retrieval of past experiences based on those predictions. The rational analysis of information scent is framed by a different assumption—that the information forager is making predictions about the expected value of different external actions—but it ends up with a derivation that parallels the rational analysis of human memory.

Fig. 5 presents an example for the purposes of discussion in this section. Fig. 5 assumes that a user has the goal of finding distal information about medical treatments for cancer and encounters a hypertext link labeled with the text that includes "cell," "patient," "dose," and "beam." The user's cognitive task is to predict the likelihood that a distal source of content contains desired information based on the proximal cues available in the hypertext link labels.

Bayes's theorem can be applied to the information foraging problem posed by situations such as those in Fig. 5. The probability that a distal content structure has desired information features, D, given a structure of proximal features, P, can be stated using Bayes's theorem as,

$$\Pr(P|D) = \Pr(D) \bullet \Pr(P|D) \tag{3}$$

where $\Pr(P|D)$ is the *posterior probability* of distal content D conditional on the occurrence of proximal structure P, $\Pr(D)$ is the *prior probability* (or *base rate*) of D, and $\Pr(P|D)$ is the *likelihood* of P occurring conditional on D. It is mathematically more tractable to conduct the analysis using *log odds*. The odds version of Bayes's theorem in Equation 3 is

$$\frac{\Pr(D \mid P)}{\Pr(D \mid \sim P)} = \frac{\Pr(D)}{\Pr(\sim D)} \cdot \frac{\Pr(P \mid D)}{\Pr(P \mid \sim D)} \tag{4}$$

where $\Pr(D|\sim P)$ is the probability of distal content D conditional on a context in which proximal structure P does not occur, $\Pr(\sim D)$ is the prior probability of D not occurring, and $\Pr(P \mid \sim D)$ is the probability of P occurring given that D does not occur.

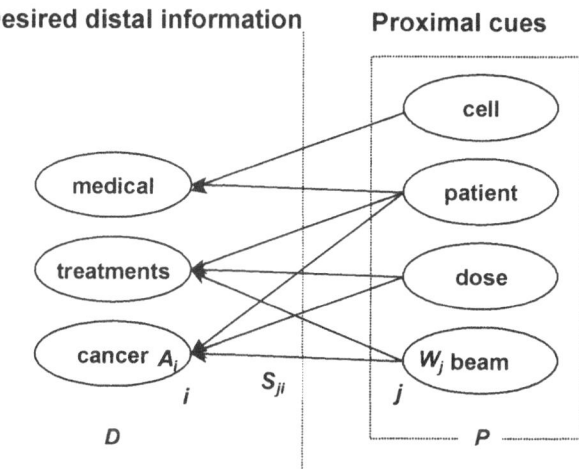

Fig. 5. A cognitive structure representing some desired goal information (D) and an encounter with some proximal information scent cues (P).

For concreteness, assume that the distal content structure D is the information contained on a Web page and that we represent D as a set of concepts corresponding to an information goal. Assume, likewise, that the proximal information scent structure P is represented as a set of the concepts corresponding to the words (images, etc.) on a Web link.

Following Anderson and Milson (1989), we make simplifying independence assumptions,[2] so that Equation 4 may be written as an odds equation for each individual feature i of the distal content structure D and each individual feature j of the proximal information scent structure,

$$\frac{\Pr(i\,|\,P)}{\Pr(\sim i\,|\,P)} = \frac{\Pr(i)}{\Pr(\sim i)} \cdot \prod_{j \in P} \frac{\Pr(i\,|\,j)}{\Pr(i\,|\sim j)} \tag{5}$$

where $\Pr(i|j)$ is the conditional probability of a distal concept or feature i occurring given that a proximal concept or feature j is present, $\Pr(\sim i|j)$ is the posterior probability of i not occurring when j is present, $\Pr(i)$ is the prior probability (base rate) of distal feature i occurring, $\Pr(\sim i)$ is the probability of i not occurring, and $\Pr(j|i)$ is the likelihood of proximal feature j occurring given distal feature i. $\Pr(i|\sim j)$ is the probability of i occurring when j does not occur. We may assume (as do Anderson & Milson, 1989) that usually the occurrence of concepts i and j together is extremely small relative to the occurrence of concept i. This means that the likelihood of concept i given that concept j is in among the information scent cues is approximately

$$\frac{\Pr(i\,|\,j)}{\Pr(i\,|\sim j)} \approx \frac{\Pr(i\,|\,j)}{\Pr(i)} \tag{6}$$

The approximation in Equation 6 is known as Pointwise Mutual Information (PMI) in the information retrieval literature (Manning & Schuetze, 1999, p. 178). As discussed later, PMI has been shown to be a very good predictor of human judgments of the similarity of word i and j.

With the approximation in Equation 6, Equation 5 may be rewritten as

$$\frac{\Pr(i\,|\,P)}{\Pr(\sim i\,|\,P)} = \frac{\Pr(i)}{\Pr(\sim i)} \cdot \prod_{j \in P} \frac{\Pr(i\,|\,j)}{\Pr(i)} \tag{7}$$

Taking the logarithms of both sides of Equation 5 leads to an additive formula,

$$\log\left[\frac{\Pr(i\,|\,j)}{\Pr(\sim i\,|\,j)}\right] = \log\left[\frac{\Pr(i)}{\Pr(\sim i)}\right] + \sum_{j} \log\left[\frac{\Pr(j\,|\,i)}{\Pr(i)}\right] \tag{8}$$

or

$$A_i = B_i + \sum_{j} S_{ji}, \tag{9}$$

where

$$A_i = \log\left[\frac{\Pr(i\mid j)}{\Pr(\sim i\mid j)}\right],$$

$$B_i = \log\left[\frac{\Pr(i)}{\Pr(\sim i)}\right],$$

$$S_{ji} = \log\left[\frac{\Pr(j\mid i)}{\Pr(i)}\right].$$

3.2.2. Mapping the Bayesian rational analysis to spreading activation

Equation 9 provides the rational grounds for a spreading activation theory of information scent. Spreading activation models are neurally inspired models that have been used for decades in simulations of human memory (e.g., Anderson, 1976; Anderson & Lebiere, 2000; Anderson & Pirolli, 1984; Quillan, 1966). In such models, activation may be interpreted metaphorically as a kind of mental energy that drives cognitive processing. Activation spreads from a set of cognitive structures that are the current focus of attention through *associations* among other cognitive structures in memory. These cognitive structures are called *chunks* (Anderson & Lebiere, 2000).

Fig. 5 presents a scenario for a spreading activation analysis. The chunks representing proximal cues are presented on the right side of Fig. 5. Fig. 5 also shows that there are associations between the goal chunks (representing needed distal information) and proximal cues (the link summary chunks). The associations among chunks come from past experience. The strength of associations reflects the degree to which proximal cues predict the occurrence of unobserved features. For instance, the word *medical* and *patient* co-occur quite frequently, and they would have a high strength of association. The stronger the associations (reflecting greater predictive strength) the greater the amount of activation flow. These association strengths are reflections of the log likelihood odds developed in Equation 9.

Previous cognitive simulations (Pirolli, 1997; Pirolli & Card, 1999) used a spreading activation model derived from the Adaptive Character of Thought–Rational (ACT–R) theory (Anderson & Lebiere, 2000). The activation of a chunk i is

$$A_i = B_i + \sum_j W_j S_{ji}. \tag{10}$$

where B_i is the base-level activation of i, S_{ji} is the association strength between an associated chunk j and chunk i, and W_j reflects attention (*source activation*) on chunk j. Note that Equation 10 reflects the log odds Equation 9 but now includes a weighting factor W that characterizes capacity limitations of human attention. One may interpret Equation 10 as reflection of a Bayesian prediction of the likelihood of one chunk in the context of other chunks. A_i in Equation 10 is interpreted as reflecting the log posterior odds that i is likely, B_i is the log prior odds of i being likely, and S_{ji} reflects the log likelihood ratios that i is likely, given that it occurs in the context of chunk j. This version of spreading activation was used in the past (Pirolli, 1997; Pirolli & Card, 1999) to develop models of information scent. The basic idea is that information scent cues in the world activate cognitive structures. Activation spreads from these cognitive structures to related structures in the spreading activation network. The amount of activation accumulating on the representation of a user's information goal provides an indicator of the likelihood that a distal source of information has desirable features based on the information scent cues immediately available to the user.

3.2.3. Relating the spread of activation to the evaluation of the utility of information foraging choices

To map this spreading activation model of information scent onto a model of rational choice (of navigation actions) involves the use of a random utility model (McFadden, 1974, 1978).[3] Random utility model (RUM) theory is grounded in classic microeconomic theory, and it has relations to psychological models of choice developed by Thurstone (1927) and Luce (1959). Its recent developments are associated with the microeconomic work of McFadden (1974, 1978).

For our purposes, a RUM consists of assumptions about (a) the characteristics of the information foragers making decisions, including their goal or goals, (b) the choice set of alternatives, (c) the proximal cues of the alternatives, and (d) a choice rule. For current purposes, we will assume a homogenous set of users with the same goal G with features, $i \in G$ (and note that there is much interesting work on RUMs for cases with heterogeneous user goals). Each choice made by a user concerns a set C of alternatives, and each alternative J is an array of displayed proximal cues, $j \in J$, for some distal information content. Each proximal cue j emits a source activation W_j. These source activations spread through associations to features i that are part of the information goal G. The activation received by each goal feature i is A_i, and the summed activation over all goal features is

$$\sum_{i \in G} A_i$$

The predicted utility $U_{J|G}$ of distal information content based on proximal cues J in the context of goal G is

$$U_{J|G} = V_{J|G} + \varepsilon_{J|G} \tag{11}$$

where

$$V_{J|G} = \sum_{i \in G} A_i$$

is the summed activation, and where $\varepsilon_{J|G}$ is a random variable error term reflecting a stochastic component of utility. Thus, the utility $U_{J|G}$ is composed of a deterministic component $V_{J/G}$ and a random component ε_{JIG}. RUM assumes utility maximization where the information forager with goal G chooses J if and only if the utility of J is greater than all the other alternatives in the choice set, i.e.,

$$U_{JIG} > U_{KIG} \text{ for all } K \in C.$$

Stated as a choice probability, this gives,

$$\Pr(J \mid G, C) = \Pr(U_{JIG} \geq U_{KIG}, \forall K \in C) \tag{12}$$

Because of the stochastic nature of the utilities U_{KIG}, it is not the case that one alternative will always be chosen over another.

The specific form of the RUM depends on assumptions concerning the nature of the random component ε_i associated with each alternative i. If the distributions of the ε_i are independent identically distributed Gumbel distributions (double exponential), then Equation 12 takes the form of a multinomial logit

$$\Pr(J \mid G, C) = \frac{e^{\mu V_{JIG}}}{\displaystyle\sum_{K \in C} e^{\mu V_{KIG}}} \tag{13}$$

where μ is a scale parameter.[4] If there is only one alternative to choose (e.g., select J or do not select J), then Equation 12 takes the form of a binomial logit,

$$\Pr(J \mid G, C) = \frac{1}{1 + e^{\mu V_{JIG}}} \tag{14}$$

For a navigation judgment, we can now specify how the computation of spreading activation yields utilities by substituting Equation 10 into Equation 11:

$$
\begin{aligned}
U_{JIG} &= V_{JIG} + \varepsilon_{JIG} \\
&= \sum_{i \in G} A_i + \varepsilon_{JIG} \\
&= \sum_{i \in G} \left(B_i + \sum_{j \in J} W_j S_{ji} \right) + \varepsilon_{JIG}
\end{aligned}
\tag{15}
$$

Equations 13, 14, and 15 provide a microeconomic model for the choice of Web links that is grounded in an underlying cognitive model of utility.

It should be emphasized that this theory of information scent is not a part of the standard ACT–R theory. The spreading activation model of information scent is not the same as the ACT–R theory of spreading activation, and the utility model based on information scent is not the same as the ACT–R model of utility. In ACT–R, spreading activation emits from goal chunks and is used to retrieve relevant chunks from memory. In the theory of information scent, cues from the external world are sources of activation; the activation levels of goal chunks are used to assess utility. In ACT–R, the utility assessments are based on the past successes and failures of actions. In the theory of information scent, the utility assessments are based on spreading activation. It is perhaps better to think of the spreading activation theory of information scent as being a kind of rational analysis of the categorization of cues according to their expected utility.

3.2.4. Estimating spreading activation strengths using PMI

As discussed in Pirolli and Card (1999), it is possible to automatically construct large spreading activation networks from on-line text corpora. In other words, we may analyze samples of the linguistic environment to provide the parameters of our cognitive models, a priori, rather than estimating those parameters, a posteriori, by fitting the models to behavioral data. The frequency of occurrence of words, and the co-occurrence frequency of words near one another can be used to estimate the base strengths, B_i, and interchunk strengths, S_{ji} in Equation 10. In the SNIF–ACT model (Pirolli & Fu, 2003) discussed later, we estimated these strengths from the Tipster document corpus (Harman, 1993) and from the Web using a program that calls on the AltaVista search engine to provide data. Recently, Turney (2001) showed that PMI scores (which approximate S_{ji}) computed from the Web can provide good fits to human word-similarity judgments.

3.3. Rational analysis of foraging in information patches

What is known as the *conventional patch model* in optimal foraging theory (Stephens & Krebs, 1986) deals with the optimal policy for the amount of time to spend in food patches before leaving. Applications of variations of that model to information foraging are discussed in Pirolli and Card (1999). Here I present another variation that has ties to work investigating stochastic models of food foraging (McNamara, 1982) and studies of the aggregate behavior of large numbers of users browsing on the Web (Huberman, Pirolli, Pitkow, & Lukose, 1998).

3.3.1. A stochastic model of patch foraging

Assume that the experiential state of the information forager at time i is represented as a state variable $\mathbf{X}_i$, and $\mathbf{X}_i = \mathbf{x}$. is a particular state value. For current purposes, assume that this state variable includes some representation of the Web page that has just been revealed and perceived by the information forager. The utility U as a function of this user state, $U(\mathbf{x})$, of continued (optimal) link browsing in this information patch can be represented by the expectation,

$$U(\mathbf{x}) = E[U|\mathbf{X}_i = \mathbf{x}] \tag{16}$$

In keeping with the previous discussion, we might assume that $U(\mathbf{x})$ is determined by choosing links having the maximum expected stochastic utility according to the RUM model. The expected time cost, t, of future (optimal) link browsing can be represented as an expectation

$$t = E[T|\mathbf{X}_i = \mathbf{x}] \tag{17}$$

where T is a random variable representing future time costs. The value $U(\mathbf{x})$ of foraging for time t in this information patch (e.g., Web site) must be balanced against the opportunity cost

$C(t)$ of foraging for that amount of time. This defines what McNamara (1982) called the *potential function, h(**x**)*, for continued foraging in this patch,

$$h(\mathbf{x}) = U(\mathbf{x}) - C(t) \qquad (18)$$

and the optimal forager is one who maximizes this potential function.

McNamara's (1982) model characterizes the opportunity cost $C(t)$ in terms of the overall long-term average rate of gain of foraging, $R*$,

$$C(t) = R * t \qquad (19)$$

In the case of information foraging on the Web, we might assume this refers to the overall average long-term rate of gain of foraging on the Web for similar tasks. The intuition behind Equations 18 and 19 is that the utility of foraging in this patch must be greater than or equal to the average rate of returns for foraging (otherwise continued foraging is incurring an opportunity cost). In other words, an information forager should continue foraging so long as,

$$U(\mathbf{x}) - R * t > 0 \qquad (20)$$

The overall average rate of gain, $R*$, could be characterized in terms of the mean utility $\bar{U}$ of going to a relevant Web site, the mean time spent setting up to go to the next relevant site (e.g., by using a search engine or guessing and typing URLs), $\bar{t}_s$, and the mean time spent foraging at the next new site, $\bar{t}$,

$$R* = \frac{\bar{U}}{\bar{t}_s + \bar{t}} \qquad (21)$$

Assuming Equation 21, we may rewrite the Inequality 20 as a rule to continue foraging so long as the rate of gain from this information patch (e.g., Web site) is greater than the expected rate of gain of going to another relevant information patch.

$$\frac{U(\mathbf{x})}{t} > R* = \frac{\bar{U}}{\bar{t}_s + \bar{t}} \qquad (22)$$

This decision to stop foraging in an information patch when the expected rate of gain drops below $R*$ is a stochastic version of the patch-leaving rule in the conventional patch model known as Charnov's marginal value theorem (Charnov, 1976), which was related to information foraging behavior in Pirolli and Card (1999). Note that the discussion here has implicitly assumed that the information forager has perfect knowledge of the relevant environmental values in Equations 19 through 22 (i.e., any learning is near asymptote). McNamara (1982) discussed how learning might be incorporated into this patch-leaving rule.

Note in Equation 22 that the average time, t, to go to a Web page that is within the same Web site as this page being visited may often be approximately the same as the time to go to a Web page at another Web site, $t_s + \bar{t}$. In such cases the decision rule to continue foraging in Equation 21 could be reduced to $U(\mathbf{x}) > \bar{U}$.

In English, this rule says

forage in an information patch until the expected potential of that patch is less than the mean expected value of going to a new patch.

In the SNIF–ACT model, it is assumed that the expected potential of a patch is estimated from the links available on Web pages, relying again on spreading activation from information scent for this assessment. SNIF–ACT also assumes that the average expected value of a new patch is estimated from past experience on the Web.

3.3.2. From individual rationality to the aggregate behavior of Web foraging: The law of surfing

One interesting consequence of this formulation of foraging in information patches is that it leads to predictions (Huberman et al., 1998) concerning patterns of aggregate behavior on the Web—specifically what has been called the *Law of Surfing*. Such aggregate distributions are of practical interest because content providers on the Web often want to know how long people will remain at their Web sites (often referred to as the *stickiness* of a Web site). More generally, the ability to relate predictions about the emergent behavior of populations from the rational models of individuals is a way of bridging psychological science and the microeconomics of the Web.

The Law of Surfing characterizes the distribution of the length, L, of sequences of page visits by Web users (see also, Baldi et al., 2003, pp. 194–199). The Law of Surfing is based on the assumption that Web users can be modeled by a random walk process in which the expected utility from continuing on to the next state is stochastically related to the expected utility of this state,

$$U(X_t) = U(X_{t-1}) + \varepsilon_t \tag{23}$$

where ε_t are independent identically distributed (IID) Gaussian distributions. Note that we are discussing aggregates of different surfers with different stochastic utility functions. The constraint of IID Gaussian noise is expected from the central limit theorem. Fig. 4, which presents the similarity of Web pages as a function of number of degrees of separation supports the assumption that the utilities of adjacent pages are related as assumed in Equation 23.

From an initial page at a Web site we expect users to continue browsing (surfing) following the random walk specified in Equation 23 until the threshold specified in Inequality 20 is reached. In the limit this is the same as the analysis of first passage times in Brownian motion. The probability density function of L, the length of sequences of Web page visits, is distributed (as are the first passage times in Brownian motion) as an inverse Gaussian distribution (Seshardri, 1993),

$$f(L) = \sqrt{\frac{\lambda}{2\pi}} L^{-3/2} e^{-\frac{\lambda}{2\mu^2 L}(L-\mu)^2} \tag{24}$$

where the parameter μ is the mean, and λ is related to the mean and variance as $\lambda = \dfrac{\mu^3}{Var[L]}$.

The inverse Gaussian is a skewed distribution that looks very much like the more familiar lognormal distribution, and it has a number of interesting properties that are discussed later in application to data. Web-site developers are often interested in the amount of content that users will visit on their Web site (or the amount of time they will spend). The Law of Surfing is relevant to characterizing these data of interest.

3.4. Summary

The general problem posed by the Web environment is one of maximizing the gain of valuable knowledge in relation to the cost of information foraging. One significant subproblem concerns the rational choice of navigation links based on available information scent cues associated with those links. Another subproblem concerns the rational allocation to time to forage in a patch before moving on to another. This section presented rational analyses of behavioral solutions to these environmental problems that will guide the SNIF–ACT cognitive model presented in the next section.

The analyses presented here extend the rational analysis approach (e.g., Anderson, 1990; Oaksford & Chater, 1998) in several ways. The first extension includes a new rational analysis of navigation in information systems based on *information scent* (Pirolli, 1997). A second extension is one of focusing rational analyses on behavior that occurs in tasks that take substantially more time than the traditional focus of rational analysis. A third extension of rational analysis has been to relate the rational analysis of individual psychology to the emergent behavior of populations.

4. Empirical evaluations

This section begins with evaluations of how well spreading activation predicts human judgments of the expected utility of browsing actions. This spreading activation model of information scent is central to the SNIF–ACT model, which is presented next along with empirical evaluations of the model. The SNIF–ACT model implements strategies that approximate the rational analyses of link choice and information-patch leaving discussed previously. This section ends with an empirical evaluation of the Law of Surfing described previously.

4.1. Information scent

Pirolli and Card (1999) presented a cognitive model, ACT–Information Foraging (ACT–IF), that simulated users in a controlled experiment in which they performed information-seeking tasks with a browser called Scatter/Gather (Cutting et al., 1992, 1993). In that controlled experiment (also reported in, Pirolli, Schank, Hearst, & Diehl, 1996), each participant was required to collect as many documents as possible relevant to a sample of tasks generated by information retrieval experts (used in the Text Retrieval Conference [TREC] workshops; Harman, 1993). ACT–IF simulated the navigation actions of these participants on the Scatter/Gather browser using spreading activation networks that were automatically computed from the base rate frequencies and co-occurrence frequencies of words in the Tipster corpus.[5]

As a side evaluation of these automatically computed spreading activation networks, Pirolli and Card (1999) compared information scent computations from the ACT–IF simulation to participant ratings of the expected utility of *cluster summaries* presented on Scatter/Gather windows. Each cluster summary was a small piece of text that Scatter/Gather used to summarize a topically related set of documents to the user, for instance:

Cluster-0 (38940) cell, patient, radiation, dose, beam, disease, treatment,

> AP: Early Results In Hospital Patient Study Sho (aid, study, percentage, health,).
> DOE: Doses of secondary radiation appearing as (radiation, dose, exposure).
> AP: Poll: AIDS Test Confidentiality Opposed in (percentage, study, drug, report).

Each such cluster summary began with a line indicting the cluster label (cluster-0), the number of documents in the cluster (38,940), and a set of keywords summarizing the central concepts for the cluster of documents (cell, patient, radiation, dose, beam, disease, treatment). These keywords were automatically computed. The next three lines summarize the three most representative documents in the cluster by their source (e.g., "AP"), some words from the title, and a list of words representing the central concepts in the document.

Each Scatter/Gather window usually presented 10 of these cluster summaries, and each set of documents summarized by a cluster summary might range from a few documents to several hundred thousand documents. As described in detail elsewhere (Pirolli et al., 1996) users interacted with the system by collecting together and decomposing clusters to get to small sets of highly relevant documents. The side study in Pirolli and Card (1999) asked users to estimate the percentage of relevant documents contained in the set of documents represented by cluster summaries presented in Scatter/Gather windows that the users encountered as they pursued their experimental tasks. The comparison of users' stated estimates to scores predicted by spreading activation in our ACT–IF simulation is presented in Fig. 6. ACT–IF used one scaling parameter used to adjust the empirical frequencies obtained from Tipster to set the spreading activation networks. The fit of ACT–IF to observed browsing behavior reported in Pirolli and Card (1999) contained no other free parameters estimated from user data. The fit to the user data in Fig. 6 uses the same spreading activation networks and a linear regression to map the activation values onto percentages.

Blackmon et al. (2002) also had success in modeling Web navigation behavior using latent semantic analysis (LSA) as a way of computing information scent, rather than using the specific spreading activation model proposed previously. Katz and Byrne (2003) also used LSA as a way of calculating information scent in a study of users' preferences for using search or browsing. LSA is a technique that provides good fits to human similarity judgments for word pairs. In application to modeling Web browsing, the assumption is that the judgments about the relevance of Web links are the same thing as judging the similarity of the Web link to a user's goal.

It appears that interchunk association strengths (S_{ij} in Equation 10) estimated by calculating PMI scores, yield similarity judgments that are comparable to LSA. Turney (2001) computed PMI scores (which approximate strength of association scores) using a standard Web search engine, by estimating $P(i|j)$ from the number of documents in which the two words i and j occurred and $P(i)$ from the number of documents in which just i occurred. On a version of the col-

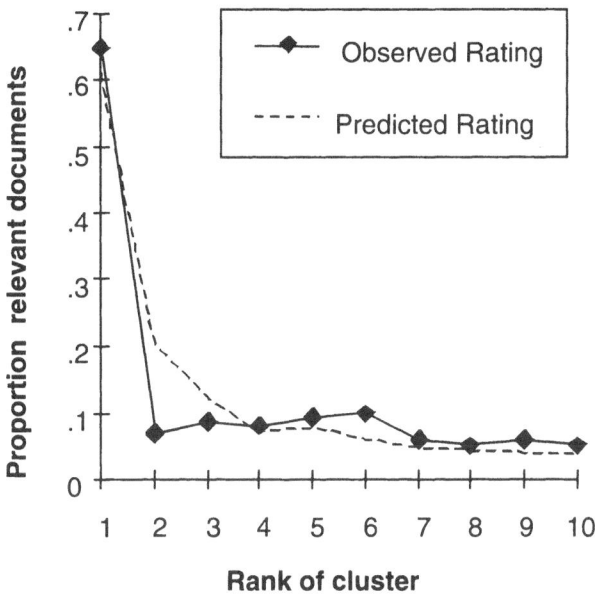

Fig. 6. Observed and predicted judgments of the expected proportion of relevant documents in a Scatter/Gather document cluster based solely on the text available in cluster summarized in Scatter/Gather windows. Observed data from $N = 24$ cluster summary ratings and predicted ratings from the ACT–IF spreading activation model of information scent operating on the same cluster summaries.

lege Test of English as a Foreign Language (TOEFL) of word similarity, students score 64.5%, LSA 64.4%, and Web-based PMI (or, equivalently, association strengths) 62.5%. Farahat, Pirolli, and Markova (2004) developed an efficient method of computing PMIs from a combination of a locally stored subset of the Web, with the Web used as a back-off for computing word similarities that are not stored locally, which scores 66.0% on the TOEFL test and outperforms LSA on other tests of human word-similarity judgments.

4.2. SNIF–ACT: Scent-based navigation and information foraging

A model called SNIF–ACT (Pirolli & Fu, 2003) was developed to simulate the participants in a Web study described in Card et al. (2001). Card et al. summarized extensive analyses of verbal protocol, WebLogger logs (which trace user and browser actions), and eye-tracking data collected from 4 participants in a Web study who worked on two tasks designed to be similar to ones reported by real-world users (Morrison et al., 2001). The tasks presented to users were

• *City task.* You are the chair of Comedic Events for Louisiana State University in Baton Rouge. Your computer has just crashed and you have lost several advertisements for upcoming events. You know that the Second City tour is coming to your theater in the spring, but you do not know the precise date. Find the date the comedy troupe is playing on your campus. Also find a photograph of the group to put on the advertisement.

• *Antz task.* After installing a state-of-the-art entertainment center in your den and replacing the furniture and carpeting, your redecorating is almost complete. All that remains to be done is to purchase a set of movie posters to hang on the walls. Find a site where you can purchase the set of four Antz movie posters depicting the princess, the hero, the best friend, and the general.

A *user-tracing architecture* (Pirolli, Fu, Reeder, & Card, 2002) was developed to match the SNIF–ACT simulation to the detailed WebLogger logs and coded verbal protocols from the 4 participants working on these two tasks.

SNIF–ACT extends the ACT–R theory and simulation environment (Anderson & Lebiere, 2000). SNIF–ACT may also be considered as an extension of the ACT–IF model of the Scatter/Gather users discussed previously (Pirolli & Card, 1999), which uses the same information scent mechanisms but different production rules. ACT–R contains three kinds of assumptions about (a) knowledge representation, (b) knowledge deployment (performance), and (c) knowledge acquisition (learning).

There are two major memory components in the ACT–R architecture: a *declarative knowledge* component and a *procedural knowledge* component. Declarative knowledge corresponds to things that we are aware we know and that can be easily described to others, such as the state of a browser, the content of Web links, the functionality of browser buttons, or information goals. Declarative knowledge is represented formally as chunks in ACT–R. Declarative chunks in ACT–R have activation values. Activation spreads from this focus of attention, including goals, through *associations* among chunks in declarative memory. Goals are also represented as chunks. At any point in time, ACT–R is focused on a single goal.

Procedural knowledge specifies how declarative knowledge is transformed into active behavior. Procedural knowledge is represented formally as condition–action pairs, or *production rules*. For instance, the SNIF–ACT simulations contain the production rule (summarized in English)

Use-Search-Engine:
IF   the goal is Goal*Start-Next-Patch
            & there is a task description
            & there is a browser
            & the browser is not at a search engine
THEN
            Set a subgoal Goal*Use-search-engine

The production (titled Use-search-engine) applies in situations where the user has a goal to go to a Web site (represented by the tag Goal*Start-Next-Patch), has processed a task description, and has a browser in front of him or her. The production rule specifies that a subgoal will be set to use a search engine. The condition (IF) side of the production rule is matched to this goal and the active chunks in declarative memory, and when a match is found, the action (THEN) side of the production rule will be executed. Roughly, the idea is that each elemental step of cognition corresponds to a production. At any point in time, a single production fires. When there is more than one match, the matching rules form a *conflict set,* and a mechanism called *conflict resolution* is used to decide which production to execute. The conflict resolution mechanism is based on a utility function. The expected utility of each matching production is calculated based on this utility function, and the one with the highest expected utility will be picked. In modeling Web users, the utility function is provided by the spreading activation model of information scent.

4.2.1. Utility and choice by information scent

In a SNIF–ACT simulation, information scent cues on a computer display activate chunks, and activation spreads through the declarative network of chunks. The amount of activation accumulating on the chunks matched by a production is used to evaluate and select productions. The activation of chunks matched by production rules is used to determine the utility of selecting those production rules. This is the most significant difference between SNIF–ACT and ACT–R, which does not have an activation-based model of utility.

For instance, the following Click-link production rule matches when a Web link description has been read:

Click-link:
IF   the goal is Goal*Process-element
          & there is a task description
          & there is a browser
          & there is a link that has been read
          & the link has a link description
THEN
          Click on the link

If selected, the rule will execute the action of clicking on the link. The chunks associated with the task description and the link description will have a certain amount of activation. That combined activation will be used to evaluate the rule. If there are two Click-link productions matching against chunks for two different links, then the production with more highly activated chunks will be selected.

The predictions made by the SNIF–ACT model were tested against the user data from Card et al. (2001). The major controlling variable in the model is the measure of information scent, which predicts two major kinds of actions: (a) which links on a Web page people will click on, and (b) when people decide to leave a site. We called the first kind of action *link-following* action, which was logged whenever a participant clicked on a link on a Web page. The second kind of action was called *site-leaving* action, which was logged whenever a participant left a Web site (and went to a different search engine or Web site). The two kinds of actions made up 72% (48% for link-following and 24% for site-leaving actions) of all the 189 actions extracted from the log files.

4.3. Link-following actions

The SNIF–ACT model was matched to the link-following actions extracted from the data sets. Each action from each participant was compared to the action chosen by the simulation model. Whenever a link-following action occurred in the user data we examined how the SNIF–ACT model ranked (using information scent) all the links on the Web page where the ac-

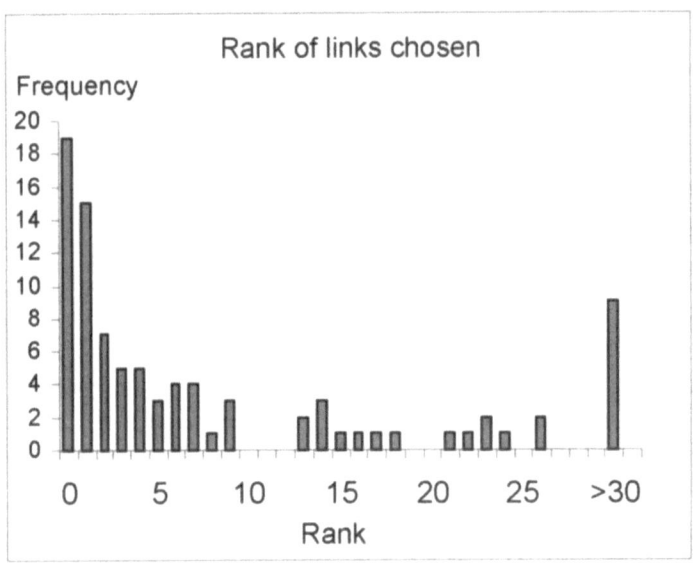

Fig. 7. Frequency that SNIF–ACT productions match link-following actions. The SNIF–ACT production rankings are computed at each simulation cycle over all links on the same Web page and all productions that match. The rankings of link-choice actions were produced by the spreading activation mechanisms for judging information scent.

tion was observed. We then compared the links chosen by the participants to the predicted link rankings of the SNIF–ACT model. If there were a purely deterministic relation between predicted information scent and link choice, then all users would be predicted to choose the highest ranked link. However, we assume that the scent-based utilities are stochastic (McFadden, 1974, 1978) and subject to some amount of variability due to users and context . Consequently we expect the probability of link choice to be highest for the links ranked with the greatest amount of scent-based utility, and that link choice probability is expected to decrease for links ranked lower on the basis of their scent-based utility values.

Fig. 7 shows that link choice is strongly related to scent-based utility values. Links ranked higher on scent-based utilities tend to get chosen over links ranked lower. There are a total of 91 link-following actions in Fig. 7. The distribution of the predicted link selection rates was significantly different from random selection $\chi^2 (30) = 18,589, p < .0001$. This result replicates a similar analysis made by Pirolli and Card (1999) concerning the ACT–IF model prediction of cluster selection in the Scatter/Gather browser. The ability of the spreading activation model of information scent in SNIF–ACT to predict link choice on the Web supports the rational analysis presented around Equation 10.

4.4. Site-leaving actions

To test how well information scent is able to predict when people will leave a site, site-leaving actions were extracted from the log files and analyzed. Site-leaving actions are defined as actions that led to a different site (e.g., when the participants used a different search engine, typed in a different URL to go to a different Web site, etc.) These data are presented in

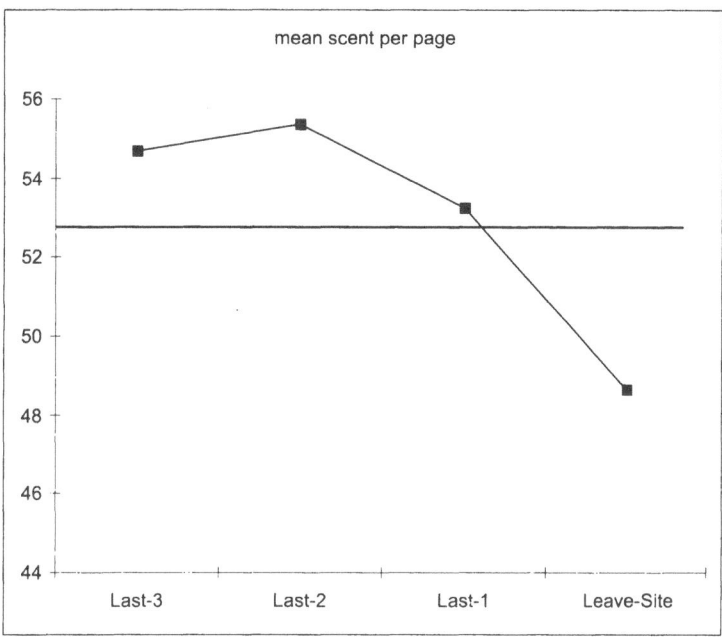

Fig. 8. Mean information scent of final four Web pages visited immediately before users left a Web site. Information scent was computed by spreading activation mechanisms. The dotted line indicates the average information scent of the first page visited on a new Web site.

Fig. 8. Each data point is the average of $N = 12$ site-leaving actions observed in the data set. The x axis indexes the four steps made prior to leaving a site (Last-3, Last-2, Last-1, Leave-Site). The y axis in Fig. 8 corresponds to the average information scent value computed by the SNIF–ACT spreading activation mechanisms. The horizontal dotted line indicates the average information scent value of the page visited by users after they left a Web site.

Fig. 8 suggests that users essentially assess the expected utility of continuing on at an information patch (i.e., a Web site) against the expected utility of switching their foraging to a new information patch. Fig. 8 suggests that spreading activation mechanisms compute activation values from information scent cues in order to reflect expected utilities associated with navigation choices. The participants in this Web study appeared to be following the patch-leaving rule developed around Equation 22.

4.5. Information patch-leaving and the Law of Surfing

SNIF–ACT is a cognitive model that approximates the rational analysis of individual Web foraging behavior. The Law of Surfing relates the rational analysis to the aggregate behavior. The Law of Surfing was tested in Huberman et al. (1998) and in addition validated in Lukose and Huberman (1998). Here, I review a couple of the evaluations in Huberman et al. to illustrate the general findings. Fig. 9 presents a typical empirical distribution of the length of paths taken by visitors to a Web site, along with a fitted inverse Gaussian distribution. Note the long positive tail, which yields a mean that is typically much larger than the mode. Fig. 10 plots an-

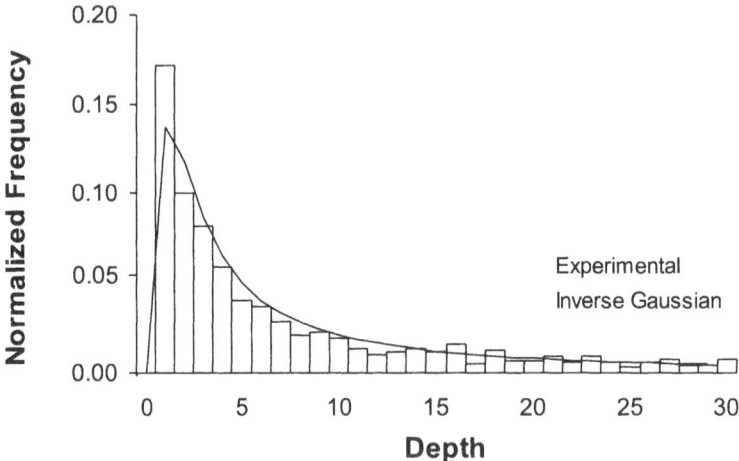

Fig. 9. The normalized frequency distribution of users as a function of depth of surfing. The observed data were collected at Boston University during late 1994 and early 1995. The fitted Inverse Gaussian distribution has a mean of μ = 51.19 and λ = 3.53.

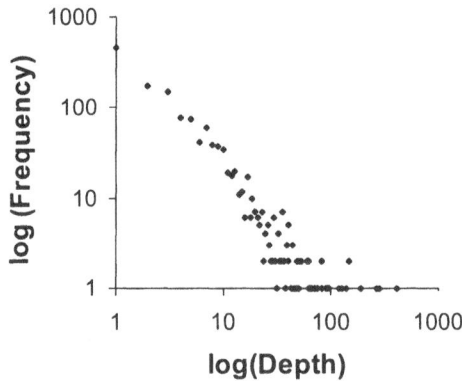

Fig. 10. The frequency distribution of surfing depths on log–log scales. Data collected from The Georgia Institute of Technology, August 1994.

other distribution of path lengths from another Web site in log-log coordinates. Figure 10 has the empirical appearance of a power-law distribution with an exponent of approximately −3/2. It turns out (Huberman et al., 1998) that this is characteristic of the inverse Gaussian distribution when $\mu < \text{Var}[L]$.

5. General discussion

Recently, Anderson (2002) drew on research in education to analyze this success of cognitive modeling. Like education, the study of information foraging provides another opportunity to develop scientific theories that simultaneously span behavioral phenomena that occur at the grain size of 100 msec to those that take 100 hr to unfold, and to produce theories and models that have practical relevance. For instance, this article integrates models of behavior associated with judgments about the relevance of individual Web links (which take approximately 1 sec) with models of Web navigation on tasks taking about .5 hr. These analyses could be integrated with "downward" models of the behavior associated with eye movements over pages, where there is already a promising rational analysis (Young, 1998), and "upward" to longer term information foraging tasks. Sense-making tasks, such as intelligence analysis (Pirolli & Card, 1999), involve trade-offs of exploration, enrichment, and exploitation over spans of hours to days to months to years that may be amenable to rational analyses and cognitive models.

Elsewhere, Anderson et al. (2004) revived the goal of "parameter-free" cognitive models (Card, Moran, & Newell, 1983). Often, cognitive models in psychology are fit to data by estimating parameters from the data themselves (e.g., through curve fitting). Information foraging models such as SNIF–ACT (Pirolli & Fu, 2003) and ACT–IF (Pirolli & Card, 1999) have achieved good fits to data using parameters set, a priori, from analysis of the information environment. The entire model of information scent and utility, save for a scaling factor estimated for ACT–IF, is determined by statistical analysis of natural language in the environment.

Theories of information foraging on the Web might be expected to improve current technology designs. The notion of information scent has already found its way into the vocabulary of Web designers (User Interface Engineering, 1999). Information foraging theory and the notion of information scent has be used to design improved search and browsing tools for the Web (Olston & Chi, 2003). Cognitive engineering models have been developed to automatically assess browser designs (Pirolli, 1998) and Web-site design (Chi et al., 2003). Given the demands in private industry and public institutions to improve the Web, and sparsity of relevant psychological theory, there is likely to be continuing demand for scientific foundations that may improve commerce and public welfare.

More generally, improvements to HII are improvements to human intelligence. This claim can be analyzed by working through Newell's (1990) discussions of knowledge and intelligence. Newell proposed that "intelligence is the ability to bring to bear all the knowledge that one has in service of one's goals" (p. 90).[6] By knowledge, Newell meant something we can attribute to a rational person that would be used to achieve his or her goals. Newell conceived of pure knowledge in a manner that transcended physical information processing limitations: In the idealized view of knowledge, everything in a body of knowledge (including all possible implications) is instantly accessible. However, people, or any physical system, can only approximate such perfect intelligent use of knowledge because the ability to bring forth the right knowledge at the right time is limited. The laws of physics limit the amount of information that can be stored or processed in a circumscribed location of space and time. Within those limits, however, intelligence increases with the ability to bring to bear the right knowledge at the right time.

Newell's (1990) discussions focused on unaided intelligent systems (people or computer programs) and the knowledge that they had available in their local memories. But there is a sense in which the world around us provides a vast external memory teeming with knowledge that can be brought forth to remedy a lack on the part of the individual. We can extend Newell's notion of intelligence and argue that intelligence is improved by enhancement of our ability to bring forth the right knowledge at the right time from the external world. Of course, the world (both physical and virtual) shapes the manner in which we can access and transform knowledge-bearing content, and thus shapes the degree to which we reason and behave intelligently. Psychological theories that provide a foundation for improved HII can provide a foundation for improving human intelligence.

Notes

1. Davison used two additional measures that yielded similar results.
2. Specifically (a) that the probability that the distal information is relevant to the user's goal is just the product of the probabilities that the individual components of the distal information are relevant to the goal, and (b) that the probability of one element i conditional on another element j is independent of all the other elements.
3. The information scent approach originally developed in Pirolli and Card (1999) is consistent with the RUM, although it was not recognized in that article.
4. Note that in both information foraging theory (Pirolli & Card, 1999) and ACT–R (Anderson & Lebiere, 2000) this equation was specified as a Boltzman equation with the substitution of $1/T$ for μ, where T is the "temperature" of the system.
5. These statistics were provided to us by our colleague Hinrich Schuetze.
6. Newell's technical definition was that "A system is *intelligent* to the degree that it approximates a knowledge-level system" (Newell, 1990, p. 90).

Acknowledgments

Portions of this research have been supported by an Office of Naval Research Contract No. N00014–96-C-0097 to P. Pirolli and S.K. Card and Advanced Research and Development Activity, Novel Intelligence from Massive Data Program Contract No. MDA904–03-C-0404 to S.K. Card and Peter Pirolli.

I would like to thank the reviewers Mark Steyvers and Michael Lee for their very constructive comments and suggestions. I also would like to thank Julie Heiser for her comments on earlier drafts. Hinrich Schuetze and Ed Chi provided technical assistance in generating data for the information scent computations.

References

Anderson, J. R. (1976). *Language, memory, and thought*. Hillsdale, NJ: Lawrence Erlbaum Associates, Inc.
Anderson, J. R. (1990). *The adaptive character of thought*. Hillsdale, NJ: Lawrence Erlbaum Associates, Inc.
Anderson, J. R. (1991). The adaptive nature of human categorization. *Psychological Review, 98*, 409–429.
Anderson, J. R. (2002). Spanning seven orders of magnitude: A challenge for cognitive modeling. *Cognitive Science, 26*, 85–112.
Anderson, J. R., Bothell, D., Byrne, M. D., Douglass, S., Lebiere, C., & Qin, Y. (2004). An integrated theory of mind. *Psychological Review, 11*, 1036–1060.

Anderson, J. R., & Lebiere, C. (2000). *The atomic components of thought.* Mahwah, NJ: Lawrence Erlbaum Associates, Inc.

Anderson, J. R., & Milson, R. (1989). Human memory: An adaptive perspective. *Psychological Review, 96,* 703–719.

Anderson, J. R., & Pirolli, P. L. (1984). Spread of activation. *Journal of Experimental Psychology: Learning, Memory, and Cognition, 10,* 791–798.

Baldi, P., Frasconi, P., & Smyth, P. (2003). *Modeling the Internet and the Web.* Chichester, England: Wiley.

Bhavnani, S. K. (2002, April). *Domain-specific search strategies for the effective retrieval of healthcare and shopping information.* Paper presented at the Conference on Human Factors and Computing Sytems, Minneapolis, MN.

Blackmon, M. H., Polson, P. G., Kitajima, M., & Lewis, C. (2002). Cognitive walkthrough for the Web. *CHI 2002, ACM Conference on Human Factors in Computing Systems, CHI Letters, 4,* 463–470.

Card, S., Pirolli, P., Van Der Wege, M., Morrison, J., Reeder, R., Schraedley, P., et al. (2001). Information scent as a driver of Web behavior graphs: Results of a protocol analysis method for web usability. *CHI 2001, ACM Conference on Human Factors in Computing Systems, CHI Letters, 3,* 498–505.

Card, S. K., Moran, T. P., & Newell, A. (1983). *The psychology of human-computer interaction.* Hillsdale, NJ: Lawrence Erlbaum Associates, Inc.

Charnov, E. L. (1976). Optimal foraging: The marginal value theorem. *Theoretical Population Biology, 9,* 129–136.

Chi, E. H., Rosien, A., Suppattanasiri, G., Williams, A., Royer, C., Chow, C., et al. (2003). The Bloodhound Project: Automating discovery of Web usability issues using the InfoScent simulator. *CHI 2003, ACM Conference on Human Factors in Computing Systems, CHI Letters, 5,* 505–512.

Cutting, D. R., Karger, D. R., & Pedersen, J. O. (1993, July). *Constant interaction-time Scatter/Gather browsing of very large document collections.* Paper presented at the Sixteenth Annual International ACM Conference on Research and Development in Information Retrieval, SIGIR 93, New York.

Cutting, D. R., Karger, D. R., Pedersen, J. O., & Tukey, J. W. (1992, June). *Scatter/gather: A cluster-based approach to browsing large document collections.* Paper presented at the Proceedings of the Fifteenth Annual International ACM Conference on Research and Development in Information Retrieval, SIGIR 92, New York.

Davison, B. (2000, July). *Topical locality in the Web.* Paper presented at the 23rd Annual International Conference on Information Retrieval, Athens, Greece.

Eiron, N., & McCurley, K. S. (2003, May). *Locality, hierarchy, and bidirectionality in the Web.* Paper presented at the Workshop on Algorithms and Models for the Web Graph, WAW 2003, Budapest, Hungary.

Farahat, A., Pirolli, P., & Markova, P. (2004). *Incremental methods for computing word pair similarity* (Tech. Rep. No. TR-04-6-2004). Palo Alto, CA: PARC.

Glimcher, P. W. (2003). *Decisions, uncertainty, and the brain: The science of neuroeconomics.* Cambridge, MA: MIT Press.

Harman, D. (1993, July). *Overview of the first text retrieval conference.* Paper presented at the 16th Annual International ACM/SIGIR Conference, Pittsburgh, PA.

Hogg, T., & Huberman, B. A. (1987). Artificial intelligence and large-scale computation: A physics perspective. *156,* 227–310.

Huberman, B. A., Pirolli, P., Pitkow, J., & Lukose, R. J. (1998). Strong regularities in World Wide Web surfing. *Science, 280,* 95–97.

Katz, M. A., & Byrne, M. D. (2003). Effects of scent and breadth on use of site-specific search on e-commerce Web sites. *ACM Transactions on Computer-Human Interaction, 10,* 198–220.

Luce, R. D. (1959). *Individual choice behavior.* New York: Wiley.

Lukose, R. M., & Huberman, B. A. (1998, October). *Surfing as a real option.* Paper presented at the International Conference on Information and Computation Economies, Charleston, NC.

Manning, C. D., & Schuetze, H. (1999). *Foundations of statistical natural language processing.* Cambridge, MA: MIT Press.

Mayr, E. (1983). How to carry out the adaptationist program? *American Naturalist, 121,* 324–334.

McFadden, D. (1974). Conditional logit analysis of qualitative choice behavior. In P. Zarembka (Ed.), *Frontiers of econometrics.* New York: Academic.

McFadden, D. (1978). Modelling the choice of residential location. In A. Karlqvist, L. Lundqvist, F. Snickars, & J. Weibull (Eds.), *Spatial interaction theory and planning models* (pp. 75–96). Cambridge, MA: Harvard University Press.

McNamara, J. (1982). Optimal patch use in a stochastic environment. *Theoretical Population Biology, 21,* 269–288.

Miller, C. S., & Remington, R. W. (2004). Modeling information navigation: Implications for information architecture. *Human Computer Interaction, 19,* 225–271.

Miller, G. A. (1983). Informavores. In F. Machlup & U. Mansfield (Eds.), *The study of information: Interdisciplinary messages* (pp. 111–113). New York: Wiley.

Morrison, J. B., Pirolli, P., & Card, S. K. (2001). A taxonomic analysis of what World Wide Web activities significantly impact people's decisions and actions. *CHI 2001, ACM Conference on Human Factors in Computing Systems, CHI Letters, 3,* 163–164.

Newell, A. (1990). *Unified theories of cognition.* Cambridge, MA: Harvard University Press.

Newell, A., & Simon, H. A. (1972). *Human problem solving.* Englewood Cliffs, NJ: Prentice Hall.

Oaksford, M., & Chater, N. (Eds.). (1998). *Rational models of cognition.* Oxford, England: Oxford University Press.

Olston, C., & Chi, E. H. (2003). ScentTrails: Integrating browsing and searching on the Web. *ACM Transactions on Computer-Human Interaction, 10,* 177–197.

Pearl, J. (1988). *Probabilistic reasoning in intelligent systems: Networks of plausible inference.* Los Altos, CA: Kaufmann.

Pirolli, P. (1997, March). *Computational models of information scent-following in a very large browsable text collection.* Paper presented at the ACM Conference on Human Factors in Computing Systems, CHI '97, Atlanta, GA.

Pirolli, P. (1998, April). *Exploring browser design trade-offs using a dynamical model of optimal information foraging.* Paper presented at the ACM Conference on Human Factors in Computing Systems, CHI '98, Los Angeles.

Pirolli, P., & Card, S. K. (1999). Information foraging. *Psychological Review, 106,* 643–675.

Pirolli, P., & Fu, W. (2003). SNIF–ACT: A model of information foraging on the World Wide Web. In P. Brusilovsky, A. Corbett, & F. de Rosis (Eds.), *User Modeling 2003, 9th International Conference, UM 2003* (Vol. 2702, pp. 45–54). Johnstown, PA: Springer-Verlag.

Pirolli, P., Fu, W., Reeder, R., & Card, S. K. (2002). A user-tracing architecture for modeling interaction with the World Wide Web. In M. D. Marsico, S. Levialdi, & L. Tarantino (Eds.), *Proceedings of the Conference on Advanced Visual Interfaces, AVI 2002* (pp. 75–83). Trento, Italy: ACM Press.

Pirolli, P., Schank, P., Hearst, M., & Diehl, C. (1996). Scatter/Gather browsing communicates the topic structure of a very large text collection. In *Proceedings of the Conference on Human Factors in Computing Systems, CHI '96* (pp. 213–220). Vancouver, BC: ACM Press.

Quillan, M. R. (1966). *Semantic memory.* Cambridge, MA: Bolt, Bernak, and Newman.

Reitman, W. R. (1964). Heuristic decision procedures, open constraints, and the structure of ill-defined problems. In M. W. Shelly & G. L. Bryan (Eds.), *Human judgements and optimality* (pp. 282–315). New York: Wiley.

Resnikoff, H. L. (1989). *The illusion of reality.* New York: Springer-Verlag.

Seshardri, V. (1993). *The inverse Gaussian distribution.* Oxford, England: Clarendon.

Simon, H. A. (1962). *The architecture of complexity.* Paper presented at the Proceedings of the American Philosophical Society, volume 106 (pp. 467–482).

Simon, H. A. (1973). The structure of ill-structured problems. *Artificial Intelligence, 4,* 181–204.

Stephens, D. W., & Krebs, J. R. (1986). *Foraging theory.* Princeton, NJ: Princeton University Press.

Stigler, G. J. (1961). The economics of information. *Journal of Political Economy, 69,* 213–225.

Thurstone, L. (1927). A law of comparative judgment. *Psychological Review, 34,* 273–286.

Tinbergen, N. (1963). On the aims and methods of ethology. *Zeitschrift für Tierpsychologie, 20,* 410–463.

Turney, P. D. (2001, September). *Mining the Web for synonyms: PMI-IR versus LSA on TOEFL.* Paper presented at the Twelfth European Conference on Machine Learning, ECML 2001, Freiburg, Germany.

User Interface Engineering. (1999). *Designing information-rich web sites.* Cambridge, MA: Author.

Winterhalder, B., & Smith, E. A. (1992). Evolutionary ecology and the social sciences. In E. A. Smith & B. Winterhalder (Eds.), *Evolutionary ecology and human behavior* (pp. 3–23). New York: de Gruyter.

Woodruff, A., Rosenholtz, R., Morrison, J. B., Faulring, A., & Pirolli, P. (2002). A comparison of the use of text summaries, plain thumbnails, and enhanced thumbnails for Web search tasks. *Journal of the American Society for Information Science and Technology, 53,* 172–185.

Young, R. M. (1998). Rational analysis of exploratory choice. In M. Oaksford & N. Chater (Eds.), *Rational models of cognition.* Oxford, England: Oxford University Press.

Cognitive Science 29 (2005) 375–419

An Activation-Based Model of Sentence Processing as Skilled Memory Retrieval

Richard L. Lewis[a], Shravan Vasishth[b]

[a]*Departments of Psychology and Linguistics, University of Michigan*
[b]*Institute for Linguistics, University of Potsdam*

Received 19 July 2004; received in revised form 2 February 2005; accepted 19 February 2005

Abstract

We present a detailed process theory of the moment-by-moment working-memory retrievals and associated control structure that subserve sentence comprehension. The theory is derived from the application of independently motivated principles of memory and cognitive skill to the specialized task of sentence parsing. The resulting theory construes sentence processing as a series of skilled associative memory retrievals modulated by similarity-based interference and fluctuating activation. The cognitive principles are formalized in computational form in the Adaptive Control of Thought–Rational (ACT–R) architecture, and our process model is realized in ACT–R. We present the results of 6 sets of simulations: 5 simulation sets provide quantitative accounts of the effects of length and structural interference on both unambiguous and garden-path structures. A final simulation set provides a graded taxonomy of double center embeddings ranging from relatively easy to extremely difficult. The explanation of center-embedding difficulty is a novel one that derives from the model's complete reliance on discriminating retrieval cues in the absence of an explicit representation of serial order information. All fits were obtained with only 1 free scaling parameter fixed across the simulations; all other parameters were ACT–R defaults. The modeling results support the hypothesis that fluctuating activation and similarity-based interference are the key factors shaping working memory in sentence processing. We contrast the theory and empirical predictions with several related accounts of sentence-processing complexity.

Keywords: Sentence processing; Working memory; ACT-R; Cognitive modeling; Interference; Decay; Activation; Parsing; Syntax; Cognitive architectures

1. Introduction

In this article we present a detailed process theory of the moment-by-moment working-memory retrievals that subserve sentence comprehension. The theory is based on general, independently motivated principles of memory and cognitive skill and provides precise quanti-

---

Requests for reprints should be sent to Richard L. Lewis, Department of Psychology, University of Michigan, 525 East University, Ann Arbor, MI 48109–1109. E-mail: rickl@umich.edu

tative accounts of reading-time data. Our vehicle for bringing sentence processing into contact with cognitive theory is Adaptive Character of Thought–Rational (ACT–R; Anderson, this issue; Anderson & Lebiere, 1998). The sentence-processing theory is derived from the application of cognitive principles, as embodied in ACT–R, to the specialized task of sentence parsing. The resulting theory construes sentence processing as a series of skilled associative memory retrievals, modulated by similarity-based interference and the fluctuation of memory trace activation. It combines insights from cognitive architectures, memory theory, and psycholinguistic theory.

By focusing on working-memory retrieval, we are departing from a long tradition in psycholinguistics and computational psycholinguistics that takes pervasive local ambiguity as both the central functional problem and the central theoretical problem in human sentence processing (e.g., Altmann & Steedman, 1988; Crocker & Brants, 2000; Ferreira & Clifton, 1986; Frazier & Rayner, 1982; Jurafsky, 1996; MacDonald, Pearlmutter, & Seidenberg, 1994; Stevenson, 1994; Tabor, Juliano, & Tanenhaus, 1998; Tanenhaus, Spivey-Knowlton, Eberhard, & Sedivy, 1995; Trueswell, Tanenhaus, & Garnsey, 1994). This has been an extremely productive line of research that continues to provide insights into cross-linguistic processing regularities (Frazier, 1998), the nature and time course of information used to resolve ambiguities, and some major architectural issues such as serial versus parallel parsing (for reviews, see Clifton & Duffy, 2001; Mitchell, 1994).

However, an even older stream of research (Chomsky & Miller, 1963; Miller & Chomsky, 1963) focuses on another major functional requirement of sentence processing: the requirement to temporarily maintain partially interpreted linguistic material so that incoming material may be integrated with it. As McElree, Foraker, and Dyer (2003) pointed out, sentences with long-distance extractions provide a clear case of this requirement, as in (1), in which the noun *toy* is interpreted as the theme of *like*:

1. This weekend we bought a toy that Amparo hoped Melissa would like.

But long-distance linguistic dependencies show up routinely in other constructions as well, as illustrated by the following extended dependency between *the dog* and the verb *stopped*.

2. The dog running around the park without his collar yesterday finally stopped barking last night.

In fact, establishing any novel relation, whether long or short, requires some memory of the immediate past. This is the functional requirement for working memory. It is not unique to language, but is a necessary feature of any computational device that must process information over time (Elman, 1990) or compute sufficiently complex functions (Newell, 1990). This requirement gives rise to the following theoretical question:

How are linguistic relations established in sentence processing—exactly what are the working-memory processes that bring prior linguistic material into contact with present material, and what are the constraints on those processes?

Most sentence-processing theories make assertions *about* composed structures or attachments (e.g., explicitly stating why one is preferred over the other or why one is computationally more costly to maintain or build than another) without explaining the basic processes and memory structures that give rise to them. (One outcome of this state of affairs is that almost no work has been done on the representation of serial order information in sentence processing, in contrast to speech production; Dell, Burger, & Svec, 1997.) This is even true for many explicit theories of working-memory resources in sentence processing, such as dependency locality theory (DLT; Gibson, 1998, 2000), which provides a characterization of the computational costs of building linguistic relations abstracted away from the processes that build them.

An important exception is the work of McElree et al. (2003) and Van Dyke and Lewis (2003), which explicitly attempts to pin down the retrieval processes in sentence comprehension. Using speed-accuracy trade-off paradigms that permit detailed time course analyses, McElree and colleagues built the case for an associative, parallel-retrieval process in parsing, and we draw extensively on this work in what follows. Van Dyke and Lewis (2003) used unambiguous and ambiguous structures to tease apart the effects of decay and interference and sketch a theory of retrieval in sentence processing that is consistent with the model presented here.

The remainder of this article is structured as follows. First, we briefly describe what it means, both theoretically and practically, to build a sentence-processing model within Adaptive Control of Thought–Rational (ACT–R) architecture. We then derive the sentence-processing theory from (a) ACT–R's architectural assumptions, (b) basic assumptions about the parsing algorithm from psycholinguistic work, and (c) representational assumptions from theoretical syntax. The next two major sections describe simulations that illustrate how the model accounts for a range of length, structural complexity, and garden-path reanalysis effects. We conclude with a summary of the theory and data coverage, provide an analysis of its relation to other accounts, and reflect on the role of ACT–R in the enterprise.

2. Toward computationally complete sentence-processing architectures

Our theoretical approach is composed of two related long-term goals. First, the sentence-processing theory should take the form of a functionally and computationally complete model of comprehension. We mean *complete* in two senses. The model should be *computationally* complete in that it specifies all the fixed mechanisms required to define any computational architecture: memories, primitive processes, and control structure (Lewis, 2000; Newell, 1973, 1990). The model should also be *functionally* complete for the particular task of real-time sentence comprehension, meaning it should specify mechanisms for lexical, syntactic, semantic, and referential processing, ambiguity resolution at all levels, and the reanalysis of ambiguous material that has been initially misinterpreted. These completeness criteria are only partially met in this model. In this article, we present a computationally complete architecture, but with an emphasis on specifying and testing the structure of working memory. Functionally, the emphasis of this model is on syntactic parsing.

The second long-term goal, alluded to previously, is to explain as much detailed psycholinguistic phenomena as possible with independent principles of cognitive processing. Cognitive architectures such as ACT–R provide the means to achieve both goals simulta-

Table 1
Summary of simulations and fits to reading-time data

Parameter	Grodner and Gibson (2005) Experiment 1	McElree et al. (2003) Experiment 2	Grodner and Gibson (2005) Experiment 2	Van Dyke and Lewis (2003) Experiment 4	Grodner, Gibson, and Tunstall (2002) Experiment 1	Double center embeddings
Decay rate	0.50	0.50	0.50	0.50	0.50	0.50
Maximum associative strength	1.5	1.5	1.5	1.5	1.5	1.5
Activation noise	0	0	0	0	0	0.3
Latency factor	0.14	0.14	0.14	0.14	0.14	0.14
Production firing time	50	50	50	50	50	50
R^2	.76	.91	.86	.77	.77	—

neously: They are computationally complete, and they explicitly embody cognitive principles in a form that can be applied to a range of tasks.

Some researchers will reject outright our second long-term goal. Seeking to provide a general cognitive basis for linguistic processes might seem ill-advised if we take language to be a specialized faculty (Chomsky, 1980; Fodor, 1983). In fact, it must be doomed to fail for two reasons under Fodor's view: Not only is linguistic processing patently not a kind of general cognition, but because nothing secure is known (or can be known) about the nature of general cognition, it is surely not the place to look for detailed insights into other mental phenomena.

However, issues of modularity and specialization are largely orthogonal to the question of whether general cognitive principles may form the basis of an explanatory theory of sentence processing. The rejection of this independence implicitly assumes that the only way that general cognitive principles can explain both Task A and Task B is if Task A and Task B both execute on the *same* general purpose mechanisms. This need not be the case, and our guiding hypothesis is that language comprehension is a specialized process operating on specialized representations that is nevertheless subject to a range of general principles and constraints on cognitive processing (Bever, 1970; Lewis, 1996; Newell, 1990). The sharing of mechanisms and resources is a separate, though important, issue.

3. Overview of the essential elements of ACT–R theory

The theoretical basis of this model is the set of components that form the core of ACT–R's theory of cognition: the declarative memory system and procedural memory system. At this stage we are not taking full advantage of ACT–R's perceptual–motor system, although we consider it a feature of the architectural approach that we may naturally extend the theory to provide models of eye movements in eye-tracking paradigms, button presses in self-paced paradigms, and so on. Our present focus is on pinning down the cognitive structure of sentence processing. For more comprehensive and up-to-date overviews of ACT–R, see Anderson (2005, this issue) and Anderson et al. (2005).

Table 2 summarizes the principles in ACT–R that we are applying to the task of sentence processing. This is not intended to be a comprehensive overview (e.g., it omits the perceptual–motor systems), and it departs from standard descriptions in some ways. We believe though it is an accurate characterization and one that highlights the aspects important for our present purposes. In what follows we briefly explain each principle.

(A1) The declarative memory component in ACT–R serves functionally as both a long-term memory (encompassing semantic and episodic memory) and a short-term working memory, although there is not a structural distinction between the two. It is useful to think of each item or chunk in declarative memory as a feature bundle that can have relations to other chunks. Fig.1 illustrates this chunk structure showing the model's representation of syntactic structure (the linguistic assumptions are discussed in more detail later).

(A2) Rather than a fixed buffer that can hold seven (Miller, 1956) or four (Cowan, 2000) chunks, ACT–R has a small set of buffers each with a capacity of one. There are three important cognitive buffers (Anderson, 2005, this issue): a *control buffer,* a *problem state buffer,* and a *retrieval buffer.* The control (or goal) buffer serves to represent current control state informa-

Table 2
The subset of ACT–R's major cognitive processing principles relevant to this model

A1 *Declarative memory of chunks.* Declarative memory consists of items (*chunks*) identified by a single symbol. Each chunk is a set of feature–value pairs; the value of a feature may be a primitive symbol or the identifier of another chunk, in which case the feature–value pair represents a relation.

A2 *Focused buffers holding single chunks.* There are an architecturally fixed set of *buffers,* each of which holds a single chunk in a distinguished state that makes it available for processing. Items outside of the buffers must be retrieved to be processed.

A3 *Activation fluctuation as a function of usage and delay.* Chunks have numeric activation values that fluctuate over time; activation reflects usage history and time-based decay. The activation affects their probability and latency of retrieval.

A4 *Associative retrieval subject to interference.* Chunks are retrieved by a content-addressed, associative retrieval process. Similarity-based retrieval interference arises as a function of retrieval cue overlap: The effectiveness of a cue is reduced as the number of items associated with the cue increases.

A5 *Procedural memory of production rules with a least-commitment, run-time control structure.* All procedural knowledge is represented as production rules (Newell, 1973)—asymmetric associations specifying *conditions* and *actions.* Conditions are patterns to match against buffer contents, and actions are taken on buffer contents. All behavior arises from production rule firing; the order of behavior is not fixed in advance but emerges in response to the dynamically changing contents of the buffers.

Note. ACT–R = Adaptive Control of Thought–Rational.

tion, and the problem state buffer represents this problem state. The retrieval buffer serves as the interface to declarative memory, holding the single chunk from the last retrieval. This structure has much in common with conceptions of working memory and short-term memory that posit an extremely limited focus of attention of one to three items, with retrieval processes required to bring items into focus for processing (McElree, 1993, 1998; McElree & Dosher, 1989; Wickelgren, Corbett, & Dosher, 1980). The state and retrieval buffers are the minimum functionally required to be able to establish novel relations between chunks. We further assume in the sentence-processing model the existence of a *lexical buffer,* which holds the results of lexical retrieval.

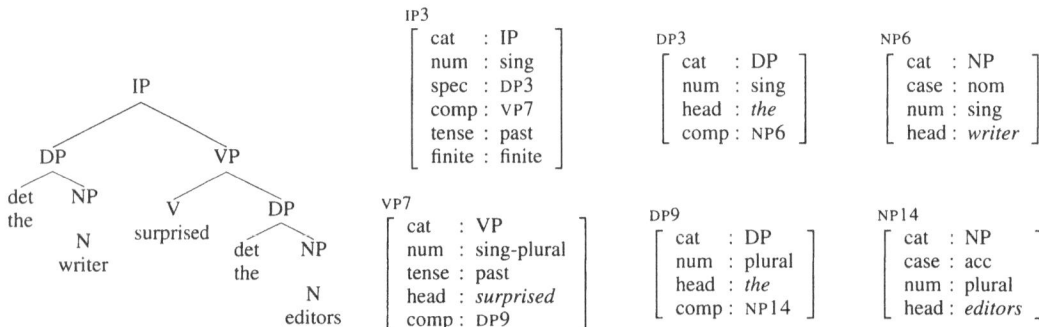

Fig. 1. An example of chunks in ACT–R's declarative memory, showing the chunk representation (right) of a syntactic structure (left).

How big is a chunk (Simon, 1974)? The claim that each buffer holds a single chunk has no empirical import if chunks can be arbitrarily defined to hold whatever information the modeler sees fit. The response to this concern is straightforward and is based on two observations. The first observation is that the answer is implicit in principle A1: A chunk is the representational element that enters into novel relations with other elements. The feature contents of two items and the novel relation between them cannot be represented in a single chunk. The second observation is that learning can of course change the representational vocabulary so that single symbols can come to denote more and more structure (Miller, 1956)—which is why we carefully restricted the first observation to *novel* relations. We take sentence comprehension to be principally a task of composing novel combinatorial representations, so the theoretical degrees of freedom in deciding what a single chunk contains are quite restricted.

(A3) All chunks have a fluctuating *activation* level, which is a function of usage history and decay. Equation 1 gives the equation for the base level activation of item i, t_j is the time since the jth retrieval of the item, and the summation is over all n retrievals.

$$B_i = \ln\left(\sum_{j=1}^{n} t_j^{-d}\right) \tag{1}$$

This equation is based on the rational analysis of Anderson and Schooler (1991) and is intended to track the log odds that an item will need to be retrieved, given its past usage history. The parameter d is estimated to be 0.5 in nearly all ACT–R models (Anderson et al., 2005), and we adopt this value. A critical feature of this equation is that it does not yield a smoothly decaying activation from the initial encoding to this time; rather, the curve has a series of spikes corresponding to the retrieval events.

The total activation of a chunk is the sum of its base activation (given in Equation 1) and an associative activation boost received from retrieval cues in the goal buffer. The activation of chunk i is defined as

$$A_i = B_i + \sum_j W_j S_{ji} \tag{2}$$

where B_i is the base activation, W_js are weights associated with elements of the goal chunk, and S_{ji}s are the strengths of association from elements j to chunk i. The total activation of a chunk determines both retrieval latency and probability of retrieval. The weights W_js are not generally free parameters in ACT–R models but are set to G/j, where j is the number of goal features, and G is the total amount of goal activation available, also set by default to 1.

Associative retrieval interference arises because the strength of association from a cue is reduced as a function of the number of items associated with the cue. This is captured by Equation 3, which reduces the maximum associative strength S by the log of the "fan" of item j, that is, the number of items associated with j.

$$S_{ji} = S - \ln(fan_j) \tag{3}$$

The final equation we require maps activation level onto retrieval latency. The latency to retrieve chunk i is given by

$$T_i = Fe^{-A_i} \qquad\qquad (4)$$

F is a scaling constant that varies across ACT–R models; in the sentence-processing model we fix F to be 0.14 and use this for all the simulations.

Principles A1 to A4 in Table 2 and associated Equations 1 to 4 together form a simple theory of associative memory retrieval specified in enough detail to make quantitative predictions. What remains to have a computationally complete framework is an answer to the following question: How are the memory retrievals organized in the service of cognition?

Principle A5 gives ACT–R's answer. Cognition is controlled by production rules (Newell, 1973)—sets of condition–action pairs. In ACT–R, all conditions are constrained to match against the contents of buffers, and all actions are constrained to make changes to buffers. The form of a typical ACT–R production is given in (3).

(3) *IF*   *control state is …*
            and chunk just retrieved has features …
            and problem state has features …
     THEN   *set new control state*
                 and update problem state
                 and set retrieval cues
                 and request retrieval

All cognitive behavior in ACT–R consists of the sequential selection and application of production rules. (See Anderson et al. (2005) for details of the ACT–R choice rule and utility learning.)

4. A theory of sentence processing based on ACT–R

Before describing and motivating the assumptions of the model in detail, it will be useful to have a high-level overview. Fig. 2 gives this overview, showing the critical buffer usage and production rule firings unfolding over time. The typical processing cycle is as follows; the numbers refer to the circled numbers in the figure.

  a. A word is attended and a lexical entry is accessed from declarative memory (1) containing syntactic information, including argument structure. The lexical entry resides in the lexical buffer (2).
  b. Based on this syntactic goal category (a kind of syntactic expectation) and the contents of the buffers, a production fires (3) that sets retrieval cues for a prior constituent to attach to.
  c. The working-memory access takes some time (4), and eventually yields a single syntactic chunk that resides in the retrieval buffer (5).

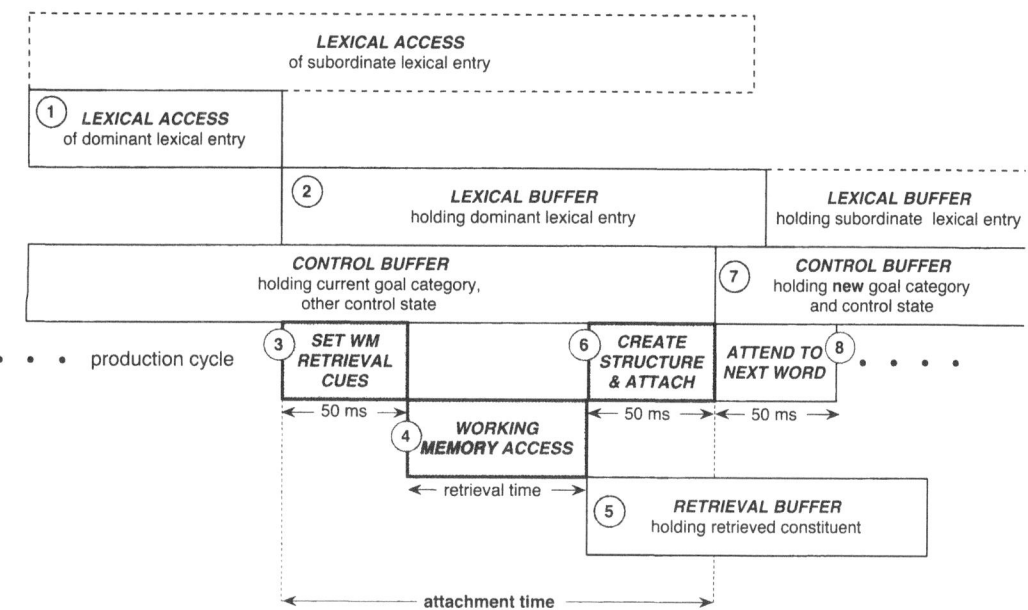

Fig. 2. Overview of the model, showing the critical focus buffers (control buffer, lexical buffer, and retrieval buffer) and processing dynamics (time flows left to right). The three key working-memory processes are shown in gray: (3) a production rule encoding grammatical knowledge sets cues for retrieval of a prior constituent; (4) a prior constituent is retrieved from working memory via parallel associative access; and (6) a second production rule creates the new structure and attaches it to the retrieved constituent.

d. Based on the retrieved constituent and lexical content, a production fires (6) that creates new syntactic structure and attaches it to the retrieved constituent. The control buffer is also updated with a new syntactic prediction (7).

e. Finally, other productions fire that guide attention to the next word.

The two production rules and retrieval processes in (3), (4), and (6) (in gray in the figure) are the critical processes of interest in this article; we refer to the time taken by all these processes jointly as the *attachment* time for a word. Apart from the new lexical buffer and parallel lexical access mechanisms, the structure of the architecture in Fig. 2 is standard ACT–R.

We now derive the details of the sentence-processing theory from a combination of ACT–R's assumptions and existing psycholinguistic evidence and theory. We first describe the major choice points in developing the model.

4.1. Major choice points in developing the sentence-processing model

Practically speaking, building an ACT–R model means specifying the contents of procedural and declarative memory. For sentence processing, there are a few immediate major choices to be made:

• How should linguistic knowledge be distributed across the procedural memory and declarative memory?

- What kind of syntactic representation should be constructed, and how does this representation map onto ACT–R chunks?
- What kind of parsing algorithm should be used (head-driven, bottom-up, etc.)?
- How is local structural ambiguity handled? Are multiple structures generated and pursued in parallel? How are misanalyses recovered from?

We take up each of these questions in turn. We acknowledge that simply asking these questions presupposes much theoretically.

4.2. Distribution of linguistic knowledge across procedural and declarative memory

We assume that the content of the lexicon resides in declarative memory. Although we are not concerned presently with modeling the details of lexical access, the immediate yield of this assumption is the prediction of context-modulated frequency effects (from Equation 1), and contextual priming effects (from Equation 3). The more contentious question concerns how grammatical knowledge is distributed across declarative and procedural memory.

Although it is impossible to construct a working model in ACT–R that consists *only* of declarative chunks, there is still a wide range of functionally viable distributions of grammatical knowledge across production rules and chunks. This range is reflected in the wide variety of parsers explored in computational linguistics and parsing theory. At one extreme are parsers with active procedures corresponding to grammar rules (e.g., traditional top-down recursive-descent parsers; Aho & Ullman, 1972). At the other extreme are parsers that encode all or most of the grammar in declarative form—the procedural component is simply an efficient interpreter of a database of declarative rules. Such parsers are widely favored in computational linguistics because they permit the easy inspection, modification, and extension of the grammar. Also closely related are lexicalist approaches to grammar and sentence processing, which seek to place as much grammatical structure as possible—perhaps all of it—into the lexicon, so that parsing is a matter of lexical retrieval and joining of retrieved structures (MacDonald et al., 1994; Schabes & Joshi, 1991; Srinivas & Joshi, 1999). Boland, Lewis, and Blodgett (2004) called this the "Full Lexical Representation Hypothesis." The potential explanatory gain, as MacDonald et al. and Jurafsky (1996) pointed out, is the unification of lexical and syntactic processing.

In contrast, the model we present here assumes that much grammatical knowledge is encoded procedurally in a large set of quite specific production rules that embody the skill of parsing. The model thus posits a structural distinction between the representation of lexical knowledge and the representation of abstract grammatical knowledge. Is such a move worth the potential loss of the unification of lexical and syntactic processing? We believe it is, for three reasons: two based on independent empirical evidence, and one based on ACT–R itself.

First, empirical evidence suggests that not all syntactic knowledge is lexicalized. Although early presentations of the lexicalist approach (e.g., MacDonald et al., 1994) emphasized the effects of traditional lexical features such as argument structure, it is now clear that a fully lexical parser must contain much more elaborated syntactic representations that go beyond argument structure. In particular, Frazier (1995, 1998) pointed out that a lexical parser must have the ability to project beyond argument structure if it is to handle adjunct attachment and phenomena in head-final constructions (Bader & Lasser, 1994; Frazier, 1987b; Hirose & Inoue, 1997; Inoue & Fodor, 1995; Konieczny, Hemforth, Scheepers, & Strube, 1997). Some highly lexicalized grammar formalisms such as LTAG have the necessary properties (Schabes & Joshi, 1991; Srinivas & Joshi, 1999), such as adjunct positions encoded in lexicalized forms. However, recent empirical work consistently points to distinctions between argument and adjunct attachment that are unexpected under the lexicalized account (Boland & Blodgett, 2001; Boland et al., 2004).

In sum, these considerations weigh against loading all the syntactic information into the declarative lexicon. At first sight it may appear that this claim is inconsistent with well-known syntactic priming effects (Bock, 1986). However, syntactic priming can be explained within the ACT–R architecture in terms of the decay of production-relevant information (in addition to decay of declarative chunks); Lovett (1998) motivated this approach in the related and more general area of choice in human perceptual and response processes. In fact, recent work on syntactic priming is consistent with this approach: (Bock & Griffin in 2000) suggested that the persistence of priming over long periods favors a long-term adjustment within a sentence-production system rather than a transient memory account.

Second, the goal of rapid processing mitigates against extra declarative retrievals in ACT–R. Even if all syntactic knowledge is not lexicalized, it still might be possible to declaratively represent and access abstract structures, as in models based on *construction grammars* (e.g., Jurafsky, 1996; McKoon & Ratcliff, 2003). However, doing so would incur extra time cost in ACT–R. In general, ACT–R theory has always assumed that skill acquisition involves a shift from declarative to procedural processing. Because we are modeling a highly practiced behavior, when the choice arises it makes sense to assume information is proceduralized, rather than remaining in declarative form.

Third, cognitive neuroscience evidence suggests that the lexicon and grammar do map onto distinct underlying declarative and procedural brain systems. The natural ACT–R mapping of lexicon–grammar to declarative–procedural is consistent with a growing body of evidence from brain imaging and patient data (Ullman, 2004; Vannest et al., 2004; Vannest, Polk, & Lewis, in press). The mapping appears to show that combinatorial processing is realized by a frontal–basal–ganglia circuit, whereas noncombinatorial lexical processing is realized by a temporal circuit. The mapping of ACT–R's declarative–procedural system to these brain areas has already been proposed on independent grounds (Anderson et al., 2005). The existence of these separate brain systems along with this independent mapping considerably lessens the concern about loss of explanatory power in moving to a dual-memory-system approach.

4.3. Declarative representation of syntactic structure

For the novel structures incrementally constructed during sentence processing, there is no choice in ACT–R for which memory system must be used. It must be the declarative system, under control of productions. Fig. 1 gives an example of the X-bar (Chomsky, 1986) structure that we assume, with a straightforward mapping onto chunks. Each chunk represents a maximal projection, with features corresponding to X-bar positions (specifier, comp, head)[1] and other syntactic features such as case and agreement.

It is important to understand that the entire syntactic tree built during the parse is not maintained in memory as a unified entity. Rather, syntactic nodes (e.g., Determiner Phrases) are maintained as chunks that are also values of features in a larger subtree. Given the architectural restrictions outlined previously, this means that accessing the information in subtrees generally requires working-memory retrievals.

4.4. Procedural representation of parsing skill

In this section we describe a simple mapping of a well-known parsing algorithm onto production rules. The resulting production rules encode both the grammar and the procedural knowledge of how to apply it.

4.4.1. Left-corner (LC) parsing

There is much evidence that real-time sentence comprehension involves incremental structure building (see references cited previously): The human parsing mechanism strives to immediately predict syntactic structure as it successively encounters each word in a sentence. In syntactic terms, this means parsing is driven by a top-down (predictive) as well as bottom-up mechanism. Johnson-Laird (1983) was the first to note that this corresponds to the well-known LC parsing algorithm (Aho & Ullman, 1972).

LC parsing can be illustrated with a simplified example. Consider the rudimentary context-free grammar of English at the top of Fig. 3.[2] We use the LC rule, defined as follows: Given an input (terminal) and a goal category, if there exists a grammar rewrite rule of which the input is a left corner, replace the input with the left-hand side (LHS) of that rewrite rule; repeat this process with the LHS nonterminal symbol until no further replacements are possible. Given this grammar, and a sentence like *the dog ran,* the parse would proceed as in Fig. 3. When the parser makes the wrong commitment, it would need to restructure the predicted tree somehow. This can be accomplished by means such as backtracking over (ranked) alternative parse trees, or repair (Lewis, 1998a). As outlined below, the ACT–R model is a repair parser.

4.4.2. The content of the production rules: How the model realizes LC parsing

Note that the parsing algorithm in Fig. 3 implicitly assumes at least two memories: a control stack of predicted categories, and a memory of the partially built structures. In the ACT–R mapping it is critical that we do not implicitly or explicitly rely on such memories; everything must be realized within ACT–R's memory systems.

The realization of LC parsing in ACT–R is straightforward. Each word triggers the sequence of events described in the previous overview and illustrated in Fig. 2. There are thus two important kinds of production rules that embody parsing skill: productions that set working-memory retrieval cues, and productions that perform attachments to retrieved constituents. The form of the two types of rules is given in (4) (compare to the general ACT–R rule schema in [3]).

(4) a. *IF   goal category is …*
             and lexical entry has features …
       THEN   set retrieval cues to …
    b. *IF   lexical entry has features …*

$$S \rightarrow NP\ VP \qquad Det \rightarrow a, the \qquad NP \rightarrow Det\ N$$
$$N \rightarrow man,\ dog \qquad V \rightarrow ran,\ saw \qquad VP \rightarrow V$$
$$VP \rightarrow V\ NP$$

INPUT: *the*
GOAL CATEGORY STACK: [S]
ACTIONS: If *the* is the left corner of any phrase structure rule then replace the stack content with the LHS of that rule. Repeat this left-corner rule until no further steps are possible. Wait for next input word. These actions yield the structure to the right:

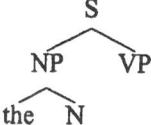

INPUT: *dog*
GOAL CATEGORY STACK: [N NP VP S]
ACTIONS: Use the left-corner rule to expand *dog* to N. Since N is predicted in the incremental structure built so far (Step 1), integrate the N built up bottom-up into the tree. Since no further applications of the left-corner rule are possible, wait for the next input.

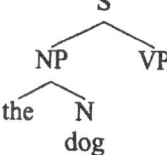

INPUT: *ran*
GOAL CATEGORY STACK: [VP S]
ACTIONS: Use the left-corner rule to expand *ran* to V, and apply this rule once again to expand to VP. Since a VP is predicted in the structure, integrate this with the tree.

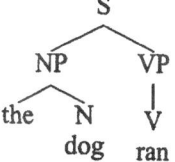

Fig. 3. A simple example of LC parsing, demonstrating the mix of bottom-up and top-down control with prediction. At top is the phrase structure grammar; the three panels and trees show the incremental parsing of *the dog ran*.

> *and retrieved constituent has features …*
> THEN   *create new constituent*
> *and attach it*

As a simple example, consider the case of projecting a determiner phrase from a determiner and attaching it to a predicted IP node (the first part of the LC example in Fig. 3). The two rules that handle this case are given as follows:

(5) a. Set-retrieval-cues-IP-goal-input-Det
        IF   goal category is IP
                and lexical entry is a DET
        THEN   set retrieval cues to IP expectation
    b. Attach-DP-as-subject-of-predicted-IP
        IF   lexical entry is category DET
                and retrieved constituent is a
                    predicted category IP
        THEN   set goal category to be NP

> and create new DP with det as head
> and attach new DP as subject
>   of predicted IP

Note that the goal category "IP" is a primitive symbol; it is not an actual constituent. Access to the IP node in the parse requires a retrieval. A useful way to think of the syntactic goal category symbols is that they form a set of control states for the sentence processor.

These rules are a compiled form of the parsing and grammatical knowledge represented in the LC algorithm plus phrase structure rules such as those in Fig. 3. There are many other specific instances of these two rule classes, but for a given phrase structure rule set (or schema set), the set is finite. The direct mapping from specific context to specific actions that the rules embody is similar to the precompiled indexing schemes used in highly efficient LC parsers (van Noord, 1997).

4.4.3. The elimination of the stack and serial order information

In the ACT–R model, there is no separate stack or chart data structure. The memory consists exclusively of the chunks representing the syntactic structure built thus far, and the only access to that memory is via the retrieval mechanisms described previously. These chunks also double as a representation of the information in the control stack; a feature "next-goal" on each constituent chunk specifies the goal category that should be pursued once the constituent is complete.

Furthermore, there is no explicit serial order representation (Lewis, 2000; cf. McElree et al., 2003). This does not mean that word order plays no role—word-order constraints are deeply embedded in the structure of all the production rules, because they depend on a distinction between the word being processed now and what has come before (what must be retrieved). The critical problem is distinguishing the relative order of two items in the past. Instead of adopting a serial order mechanism, we are initially pursuing an extreme hypothesis about serial order representation in the human parser: There is none. Instead, the processor relies on the ability of retrieval cues to discriminate candidate attachment sites, and in cases where retrieval cues cannot discriminate, the processor relies (implicitly) on activation level. We explore the implications of this assumption later in the simulations of center embeddings.

4.4.4. Retrieval cues for embedded structures and gapped structures

One virtue of using a top-down goal category to guide behavior is that the processor can be sensitive to whether or not it is processing an embedded clausal structure or gapped structure (such as a relative clause). For gapped structures this is useful because it provides the triggering cue to attempt retrieval of the dislocated element. (We assume here a syntactic representation that uses empty categories, although the model could be reformulated as a direct association model.) Thus, in addition to goal symbols such as VP and IP there are also corresponding goal symbols VP-gapped and IP-gapped, and gap features on the constituent chunks themselves. Similarly, the model is sensitive to whether or not it is parsing an embedded clause by using goal symbols such as IP-embedded and VP-embedded and corresponding embedded features on the constituents themselves. This helps the processor functionally to discriminate between the predicted main and predicted embedded clauses during retrieval.

These assumptions have precedents in both the linguistic and psycholinguistic literature. The passing down of gap features (Gazdar, 1981; Gazdar, Klein, Pullum, & Sag, 1985) is similar to the use of the SLASH feature in head-driven phrase-structure grammar (Pollard & Sag, 1994), and the vocabulary of gapped categories is inspired by (combinatorial) categorial grammar (Bar-Hillel, 1953; Steedman, 1996). The distinction between embedded and main clauses is independently needed because these structures admit different syntactic phenomena (Green, 1976; Hooper & Thompson, 1973; Koster, 1978).

4.5. *Responding to local ambiguity: Serial, probabilistic, repair parsing*

To fully specify a parser we must specify mechanisms for handling local lexical and structural ambiguity. In this model, these mechanisms follow from the ACT–R architecture and implicitly provide a theory of ambiguity resolution. Because our current focus is on working-memory retrieval and our initial set of results can be described independently of ambiguity resolution, the following brief summary will suffice:

- Lexical access proceeds via ordered access modulated by frequency and context, with competition effects. This is a direct result of Equations 1, 2, and 3.
- Structural ambiguity is resolved probabilistically via a combination of working-memory factors (such as recency) and ACT–R's rational production choice rule. This is a direct result of ACT–R's control structure and the declarative memory equations.
- A single structural interpretation is pursued, although multiple possibilities are locally generated in parallel. The single-path nature of the system is indirectly a consequence of associative retrieval interference, which mitigates against maintaining multiple similar structures.
- Limited recovery from misanalyses is initiated opportunistically by the reactivation of discarded structures, and completed by simple repair. This is indirectly a consequence of the architecture as well; it is the most natural outcome, given the serial nature of the parse. The discarded structures are simply chunks in memory like any other, but they have not participated in the parse and thus have not been retrieved since their initial creation.

4.6. *An example parse*

We now briefly illustrate the behavior of the model on a simple example that brings all the mechanisms together. Consider the objective-relative (OR) clause structure in (6):

(6) The lawyer that the editor hired admired the writer.

Fig. 4 gives a partial trace of the model parsing this example, highlighting the processing of four words: the initial *the,* the complementizer *that,* and the embedded and main verbs. Each line prefixed by a timestamp indicates either the completion of a production rule firing or a memory retrieval. The two productions for *the* are the same ones previously given in (4).

The total syntactic attachment time indicated in the trace is the sum of the processing time after lexical access until the attachment is complete (see also Fig. 2). As described previously, each word is typically processed by (a) a lexical access (not shown in the trace), (b) a produc-

```
READING: "the" at  0.235


  0.286: Set-Retrieval-Cues-IP-Goal-Input-Det

  0.338: IP11 Retrieved

  0.388: Attach-DP-As-Subject-Of-Predicted-IP


  ================================================
  Attachment time for THE:  0.152
  ================================================


READING: "lawyer" at  0.623
READING: "that" at  1.016


  1.067: Set-Retrieval-Cues-Input-Comp

  1.103: DP12 Retrieved

  1.153: Attach-CP-As-Modifier-Of-Retrieved-DP


  ================================================
  Attachment time for THAT:  0.136
  ================================================


READING: "the" at  1.388
READING: "editor" at  1.947
READING: "hired" at  2.356


  2.408: Set-Retrieval-Cues-Goal-VP-Gapped-Input-V

  2.441: IP16 Retrieved

  2.491: Attach-VP-Transitive-DP-Object-Gap

  2.548: DP16 Retrieved

  2.598: Attach-Filler-In-Object-Gap


  ================================================
  Attachment time for HIRED:  0.240
  ================================================


READING: "admired" at  2.833


  2.885: Set-Retrieval-Cues-Goal-VP-Input-V-Finite

  2.963: IP11 Retrieved

  3.013: Attach-VP-Transitive-DP-No-Gap
  ================================================
  Attachment time for ADMIRED:  0.178
  ================================================
```

Fig. 4. A partial trace generated by the model in processing the sentence *The lawyer that the editor hired admired the writer*. Times are in seconds; see text for an explanation of "attachment time."

tion rule to set retrieval cues, (c) a working-memory retrieval, and (d) a production rule to perform the attachment.

When *the* is read at time 0.235, retrieval cues are set (via a production rule) for a predicted IP given the input determiner; the production rule execution takes 50 msec (the ACT–R default). The retrieval cues trigger a retrieval of the IP, which takes 53 msec. Finally, a second production rule attaches the determiner phrase as subject of the predicted IP, and this also takes another 50 msec. Adding 50, 52, and 50 gives the total attachment time for *the*: 152 msec.

The processing of *that* is similar. When it is read at time 1.067, retrieval cues are set for a DP, and the previously seen DP is retrieved and the CP attached to it as a modifier. The two productions take 50 msec each, and the retrieval 36 msec, resulting in a total attachment time of 136 msec for *that*.

For *admired* the attachment time 178 msec is the sum of 50 msec, 78 msec, and 50 msec (the two 50 msec for production rule execution times, and 78 msec for retrieving the IP node, which is expecting the VP). For *hired,* there is an additional production firing and retrieval associated with retrieving the relative pronoun and attaching a coindexed trace to fill the gap in the relative clause structure.

4.7. Summary of the major theoretical claims about working memory

Before describing the quantitative simulations, let us take stock and summarize the major theoretical claims about working memory in sentence processing. Table 3 lists these claims. It is important to understand that these are simply instantiations of the basic ACT–R principles listed in Table 2, and formalized in part by Equations 1 to 3.

Table 3
Summary of the major theoretical claims about working memory in sentence processing

---

SP1 *Declarative memory for long-term lexical and novel linguistic structure.* Both the intermediate structures built during sentence processing and long-term lexical content are represented in declarative form by chunks, which are bundles of feature–value pairs.

SP2 *Extremely limited working-memory focus.* Single-chunk buffers hold (a) the results of lexical access, (b) the constituent just retrieved from working memory, and (c) local control state, including the syntactic goal category. Active processing is restricted to the contents of these buffers.

SP3 *Activation fluctuation as a function of usage and delay.* Chunks representing constituents during sentence processing and lexical entries in long-term memory have activation values that fluctuate over time; activation reflects usage history and time-based decay. The activation affects their probability and latency of retrieval.

SP4 *Associative retrieval subject to interference.* All chunks, including chunks comprising working memory, are accessed by a content-addressed, associative retrieval process. Retrieval cues are a subset of the target chunk's features. Similarity-based retrieval interference arises as a function of cue overlap: The effectiveness of a cue is reduced as the number of items associated with the cue increases.

SP5 *Efficient parsing skill in a procedural memory of production rules.* A large set of highly specific production rules constitute the skill of parsing and a compiled form of grammatical knowledge. The parsing algorithm is best described as a LC algorithm, with a mix of bottom-up and top-down control. Sentence processing consists of a series of memory retrievals guided by the production rules realizing this parsing algorithm.

---

5. Quantitative effects of activation fluctuation and interference

In this section we describe a series of five sets of simulations that test the ability of the model to explain relatively complex patterns of reading times and provide quantitative accounts of those patterns. We are primarily interested here in the effects of decay and interference on syntactic processing, so we focus on experimental contrasts that factor out effects of lexical and other processing. These provide the most straightforward test of the model's working-memory assumptions.

5.1. How the fits were obtained

In all the simulations reported in the following sections, the ACT–R model parses, word-by-word, examples of the sentences presented in the actual experiments. There are four important quantitative parameters that affect the predictions: the decay rate b (Equation 1), the maximum associative strength S (Equation 3), the latency factor F (Equation 4), and the production execution latency. The decay rate b has a standard ACT–R value of 0.5. The production execution latency has a standard ACT–R value of 50 msec, which is also consistent with the estimate in EPIC (Meyer & Kieras, 1997) and SOAR (Newell, 1990). The associative strength S is estimated at 1.5 in several models of memory interference experiments (Anderson & Reder, 1999; Anderson et al., 2005), and we adopt that value here. We estimated the latency factor at 0.14; this just serves as a scaling factor. In sum, we are simply using ACT–R default parameter ranges and estimating a single parameter where the theory does not provide explicit guidance.

All of the simulations described here were fitted with the same parameter values; there is no parameter variation across simulations. Furthermore, the plotted "fits" are not standard linear regression fits with two parameters (a slope and intercept). Because ACT–R has its own scaling parameter, it is not appropriate to further scale the values with another regression parameter. Thus, we plot here absolute differences against the data, with a constant adjustment to factor out button-press times, and so forth.

Table 1 summarizes the simulation, the parameters, and the R^2 values. We report these R^2 values as a rough summary of how well the model is capturing the data patterns, with the caveat that correlations are based on a two-parameter linear model, and we are interested here in assessing the match of absolute time differences across the simulations. Rather than rely on a goodness of fit summarized by a single quantity, we prefer instead to guide the reader through analyses of the data that focus on individual contrasts of theoretical interest.

5.2. Simulation 1: Subject relatives (SRs) versus ORs

There is considerable psycholinguistic evidence that English OR clauses are more difficult than SRs (e.g., Ford, 1983; Hakes, Evans, & Brannon, 1976; Holmes & O'Regan, 1981; Hudgins & Cullinan, 1978; King & Just, 1991; Larkin & Burns, 1977): They are processed more slowly and often result in poorer comprehension. Consider the pair in (7), from King and Just (1991), in which each relative clause modifies the subject of the main clause.

(7) a. [SR] The reporter who attacked the senator admitted the error.
    b. [OR] The reporter who the senator attacked admitted the error.

There have been many explanations proposed for this contrast. Most explanations can be classified into one of four types: (a) *distance* explanations (Gibson, 1998; Grodner & Gibson, in press; Just & Carpenter, 1992) attribute the contrast to the fact that in the OR structure there is a greater distance between the embedded verb *attacked* and the relative pronoun; (b) *double function* explanations (Bever, 1970; Sheldon, 1974) attribute the contrast to the fact that in the OR clause, one noun phrase (*reporter*) simultaneously plays two different underlying functions (object of the embedded clause and subject of the main clause); (c) *experience-based* explanations (MacDonald & Christiansen, 2002) attribute the contrast to the putative divergence of the OR structure from the canonical and much more frequent subject–verb–object English word order, or to the relatively rare occurrence of ORs (Korthals, 2001); and (d) *reanalysis* explanations attribute the contrast to the local ambiguity at *who* and the preference to initially analyze the structure as a SR (Frazier, 1987a, 1989).

There are empirical and conceptual difficulties with the latter three explanations; see Grodner and Gibson, in press, for a critique of the reanalysis and experience explanations.[3] One problem with the double-function account is that it does not hold up cross-linguistically. Studies in Japanese (Hakuta, 1981) and Hebrew (Schlesinger, 1975) have demonstrated that it is not double function, per se, but center embedding, or amount of interruption, that best accounts for difficulty.

We are interested in accounting for the pattern of reading times at embedded verbs and main verbs in both structures. King and Just (1991) collected word-by-word self-paced reading times for structures such as (7), but Grodner and Gibson (in press) pointed out that reading times for the main verb may be inflated in the OR case due to spillover from the embedded verb, so it is difficult to get a clean comparison of processing time on the main verb between the two structures. They corrected this (and other problems) by using the following structures:

(8) a. [SR] The reporter who sent the photographer to the editor hoped for a story.
   b. [OR] The reporter who the photographer sent to the editor hoped for a story.

Their observed reading times and the model generated times are shown in Fig. 5. The model captures the basic contrast between SRs and ORs; the highest reading times are on the embedded verb in the OR. There are two further interesting patterns present in the data. First, the effect of extraction type is greatest at the embedded verb. Table 4 shows this contrast. In fact, in the human data, there is no effect at all of extraction type on the main verb; there is a small effect present in the model. Second, the reading times on the main verb are shorter than the embedded verb in the OR condition, but longer in the SR condition.

This same interaction of extraction type and verb has been observed in two earlier reading-time studies. The pattern was observed in the King and Just (1991) experiment for all but the low working-memory-span subjects—despite the possibility of spillover inflating the reading times on the main verb as discussed previously. The pattern was also observed in a self-paced paradigm, using centrally presented words (Gordon, Hendrick, & Johnson, 2001). Neither the CC-READER model (Just & Carpenter, 1992) nor the locality-based integration metric (Grodner & Gibson, in press) captures this aspect of the data. The locality account predicts the highest reading times at the main verb in both constructions; the CC-READER model predicts higher times for the SR construction and equal times for the OR construction.

The contrasts between the SR embedded and main and OR main verbs in the ACT–R model are due only to differences in working-memory retrieval times, modulated by activation fluctu-

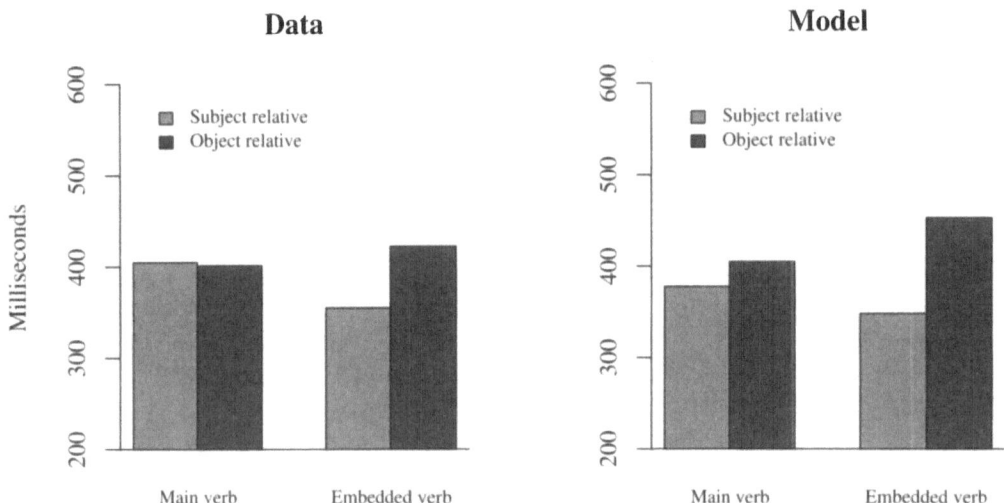

Fig. 5. Model-generated and observed self-paced reading times on embedded verbs and main verbs in SR and OR clauses. Data from Experiment 1, Grodner and Gibson (2005).

ation. The extra time at the object-gap filling is due in part to an extra retrieval and production cycle, as explained previously. The subject gap is posited and filled at the relative pronoun *who*, so an additional retrieval is not required at the verb.[4] Table 4 suggests that the model is overestimating the cost of the object-gap filling for this experiment. It is possible to adjust the latency factor to reduce the mismatch considerably, but we prefer for now to explore the implications of a consistent set of parameters across all of our simulations, until we adopt a principled basis for modeling individual differences.

5.3. Simulation 2: Estimating the time for working-memory retrieval

McElree et al. (2003) applied McElree's (1993, 1998) theory of short-term memory to sentence processing. The major assumptions are consistent with the ACT–R model: There is a very limited focus of attention, and items outside the focus must be retrieved. McElree et al. (2003) tested this idea in sentence processing using a speed–accuracy–trade-off (SAT) paradigm designed to provide an assessment of the cost of retrieving an element outside the focus. Participants read sentences in a rapid serial visual presentation paradigm at 250 msec per word

Table 4
Contrasts in reading times (msec) for Subject Relatives and Object Realtives (data from Experiment 1, Grodner & Gibson, 2005)

Contrast	Model	Data
Object Relative vs. Subject Relative		
At main verb	27	−4
At embedded verb	105	67
Interaction of verb and extraction	78	71

and responded to a tone at various time lags after the onset of the final word of the sentence. The task was to perform a semantic acceptability judgment, which required the participants to successfully parse the sentence. Examples of their materials and conditions are given in (9):

(9) a. [None] The book ripped.
  b. [OR] The book that the editor admired ripped.
  c. [PP–OR] The book from the prestigious press that the editor admired ripped.
  d. [OR–SR] The book that the editor who quit the journal admired ripped.
  e. [OR–OR] The book that the editor who the receptionist married admired ripped.

The four conditions manipulate the amount and type of interpolated material between the main verb *ripped* and the subject *book* (None = no interpolation, PP-OR = prepositional phrase and OR, OR–SR = subject relative embedded within OR, OR–OR = double embedded OR). Their SAT analysis provides an estimate of the time to retrieve the displaced subject NP at the main verb.

The analysis revealed three distinct retrieval times: one fast time for the None condition, an intermediate time for the OR, PP–OR, and OR–SR conditions, and a third longer time for the OR–OR condition. Note that the OR–OR construction is the notoriously difficult classic double center embedding, and McElree et al. (2003) suggested that the inflated OR–OR times result from additional recovery times from misparsing this construction and do not provide a clean estimate of retrieval times. Fig. 6 shows the results, without the OR–OR condition. (We further discuss center embeddings later in the article.)

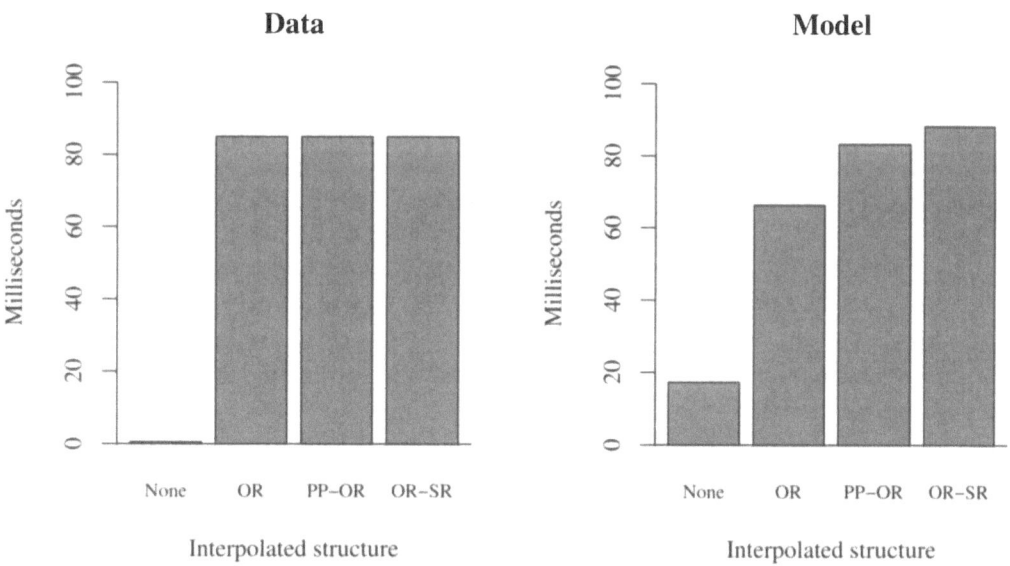

Fig. 6. Model-generated and observed processing times on the final verb as a function of interpolated structures. Data from Experiment 2, McElree et al. (2003). None = no interpolation, OR = interpolated object relative, OR-PP = interpolated prepositional phrase and object relative, OR-SR = interpolated subject relative embedded within object relative.

Table 5
Contrast in processing times (msec) at final verb (data from Experiment 2, McElree et al., 2003)

Contrast	Model	Data Estimate
Distal vs. local attachment	62	85
PP–OR vs. OR	17	0
OR–SR vs. PP–OR	5	0

McElree et al. (2003) argued that the 85-msec difference in processing dynamics between the None and OR/PP–OR/OR–SR conditions provides an estimate of the time required for the working-memory retrieval of the subject NP. The ACT–R model differs somewhat from the structural assumptions of the McElree et al. (2003) model, in that there is still a retrieval required in the None condition. Nevertheless, the attachment time in the None condition is much faster (by 62 msec) than the other three conditions. We characterize the contrast as "distal versus local" rather than "retrieval versus no retrieval."

Perhaps the most striking aspect of the McElree et al. (2003) data is the lack of difference between the PP–OR, OR–SR, and OR conditions. Table 5 summarizes this effect and shows that the ACT–R model captures this basic interaction, though still predicting some minor differences.

5.4. Simulation 3: Effects of interpolated material on main verbs and embedded verbs

Grodner and Gibson (in press) conducted a follow-up to their relative clause experiment that lets us further test the model's predictions concerning the effects of interpolated material on main verbs (as in McElree et al., 2003) and embedded verbs. Examples of their materials are given in (10)

(10) a. [main:None] The nurse *supervised* the administrator while …
      b. [main:PP] The nurse from the clinic *supervised* the administrator …
      c. [main:RC] The nurse who was from the clinic *supervised* the administrator while …
      d. [embedded:None] The administrator who the nurse *supervised* scolded the medic while …
      e. [embedded:PP] The administrator who the nurse from the clinic *supervised* scolded the medic while …
      f. [embedded:RC] The administrator who the nurse who was from the clinic *supervised* scolded the medic while …

The conditions contrast the effect of modifying the subject of either the main verb or the embedded verb: The conditions None (no modification), PP modification, and RC modification are crossed with main verb and embedded verb (*supervised* in the previous example). These structures thus provide another test of both how the model processes relative clauses (the main vs. embedded contrast) and how the model handles nonlocal attachments across different structures (the modifier type contrasts).

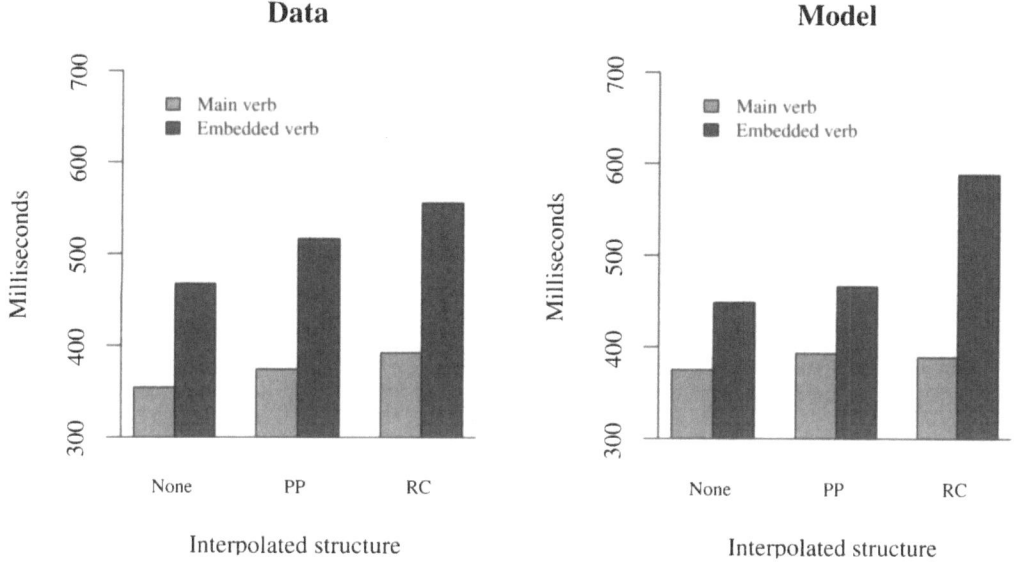

Fig. 7. Model-generated and observed self-paced reading times on main verbs and embedded verbs as a function of interpolated structure (different modifiers of the verb's subject). Data from Experiment 2, Grodner and Gibson (2005). None = no interpolated structure, PP = prepositional phrase, RC = subject relative clause.

Fig. 7 shows the experimental data and the model-generated reading times. There are a couple of important contrasts to focus on; Table 6 summarizes these. First is the effect of embedding for the OR: a general increase in reading times for the embedded verbs versus the main verbs, as discussed previously. Second, there is an effect of interpolated structure (modifier type), with the relative clause modification showing the largest effect. Furthermore, the main verb reading times are remarkably flat, which is consistent with the result of McElree et al. (2003) discussed previously. The ACT–R model also predicts an interaction of verb type and modification as Fig. 7 and Table 6 reveal. The interaction arises for two reasons: There is increased retrieval interference at the embedded verb (because the expectation for the main verb overlaps in retrieval cues), and the effect of decay is felt on both the retrieval of the expected IP

Table 6
Contrasts in reading times (msec) for main verbs and embedded verbs as a function of subject modification (data from Experiment 2, Grodner & Gibson, 2005)

Contrast	Model	Data
Embedded vs. main	139	115
Distal vs. local	49	48
PP modifier vs. none		
Main verb	20	18
Embedded verb	49	18
RC modifier vs. none		
Main verb	38	14
Embedded verb	88	140

node and the retrieval of the object-gap filler. This prediction is consistent with the human data, although the interaction was not statistically significant.

5.5. Simulation 4: Effects of length and interference in garden paths and their unambiguous controls

The experiments we have considered thus far provide good quantitative tests of the model's account of distal versus local attachments under various conditions, but do not yet begin to clearly distinguish the effects of decay and interference, both of which are fundamental parts of ACT–R's (and thus the model's) theory of memory retrieval. Van Dyke and Lewis (2003) recently used locally ambiguous garden-path sentences in addition to unambiguous structures in an attempt to tease apart the effects.

5.5.1. Materials and design

Examples of the materials are given in (11). There were two manipulations: a structural interference ([int]) and distance ([short]) manipulation, crossed with an ambiguity ([(un)ambig]) manipulation.

(11) a. [unambig:short] The assistant forgot that the student *was standing* in the hallway.
  b. [unambig:low-int] The assistant forgot that the student who was waiting for the exam *was standing* in the hallway.
  c. [unambig:high-int] The assistant forgot that the student who knew that the exam was important *was standing* in the hallway.
  d. [ambig:short The assistant forgot the student *was standing* in the hallway.
  e. [ambig:low-int The assistant forgot the student who was waiting for the exam *was standing* in the hallway.
  f. [ambig:high-int The assistant forgot the student who knew that the exam was important *was standing* in the hallway.

The critical region of interest is the same in all conditions: The reading time on the embedded verb *was standing*. Note that the embedded clause is a sentential complement, not a relative clause in this case. We are interested in the effect of material interpolated between the subject NP *the student* and the embedded verb.

Consider the unambiguous conditions first. In the short condition, nothing intervenes between the subject NP and the verb. In the low-interference condition, there is an intervening SR clause with a prepositional phrase. The high-interference condition is equally long, but instead of a prepositional phrase, it includes another intervening sentential complement, which provides additional structural interference. Van Dyke and Lewis (2003) argued that the contrast between the short- and low-interference conditions provides an estimate of a *distance* effect, whereas the contrast between the low-interference and high-interference conditions provides an estimate of an *interference* effect.

Now consider the ambiguous conditions. Van Dyke and Lewis (2003) reasoned that garden-path structures will show *additional* distance effects, but *no additional* interference effects. The argument is as follows. Distance effects should show up most clearly when memory traces are left to decay without intervening retrievals and processing. A garden-path sentence with a clear preference for one structure over another provides just such a situation: The dispreferred structure is left to decay without additional processing during the ambiguous region (under a serial parsing account). However, the interference effect should be about the same, because the intervening structures in the ambiguous region are the same.

To clear: What is at issue here is the effect of distance and interference on the garden-path effect itself. The prediction is not that garden-path structures will show no interference effects; rather, relative to their unambiguous controls, there is no *additional* interference. In contrast, the prediction for distance is the opposite. Relative to their unambiguous controls, garden-path structures *will* show an additional distance effect.

The ambiguous materials in (11) use the classic subject–object garden path (e.g., Ferreira & Henderson, 1991) with lexical material carefully chosen to bias toward the simple object NP reading (*the student* is initially taken to be the object of *forgot,* rather than the subject of an up-coming embedded clause). Thus, the critical region *was standing* in the ambiguous conditions now serves to disambiguate toward the sentential complement structure, and it is here that we expect the garden-path effect to arise (Ferreira & Henderson, 1991).

5.5.2. Empirical and modeling results

The results of Van Dyke and Lewis (2003)'s self-paced reading study confirm these predictions: There was an effect of interference on the unambiguous constructions but almost no effect of interference on the garden-path effect. In contrast, there was no distance effect on the unambiguous constructions, but there was a significant distance effect on the garden-path effect—a clear cross-over interaction between effects of distance and interaction on reanalysis and attachment.

In the ACT–R model, both the noun phrase and sentential complement structures are generated at the locally ambiguous verb *forgot,* but lexical access delivers the NP complement argument structure first, so it is pursued in the parse (realizing a simple race model of ambiguity resolution), whereas the other structure continues to decay without further processing. At the critical disambiguating verb, this discarded structure must be retrieved.

Fig. 8 shows the human times and the model-generated times at the critical verb for the six conditions of the experiment (the times are residuals after regressing out length and word position effects). (The figure also shows times for another region in the unambiguous constructions; we return to this data in the discussion of syntactic load effects).

This is a fairly complex pattern of data, and there are several important contrasts. The contrasts are summarized in Table 7. First, there is a main effect of ambiguity—a garden-path effect of 61 msec. The model predicts this effect because of the additional retrieval time taken in the ambiguous case to reactive the discarded memory item. Second, there is also a main effect of interference (comparing the high- and low-interference conditions); the model predicts this because of additional retrieval interference from the intervening IP node in the high-interference structure.

Looking in more detail at the effects of interference, the model correctly predicts a greater effect of interference on attachment than on reanalysis. Looking in more detail at the effects of length, the model correctly predicts that length effects are greater than interference effects for reanalysis. This is because in the unambiguous case the pursued structure has one additional retrieval as a result of participating in the parse, and this reduces the effects of decay.

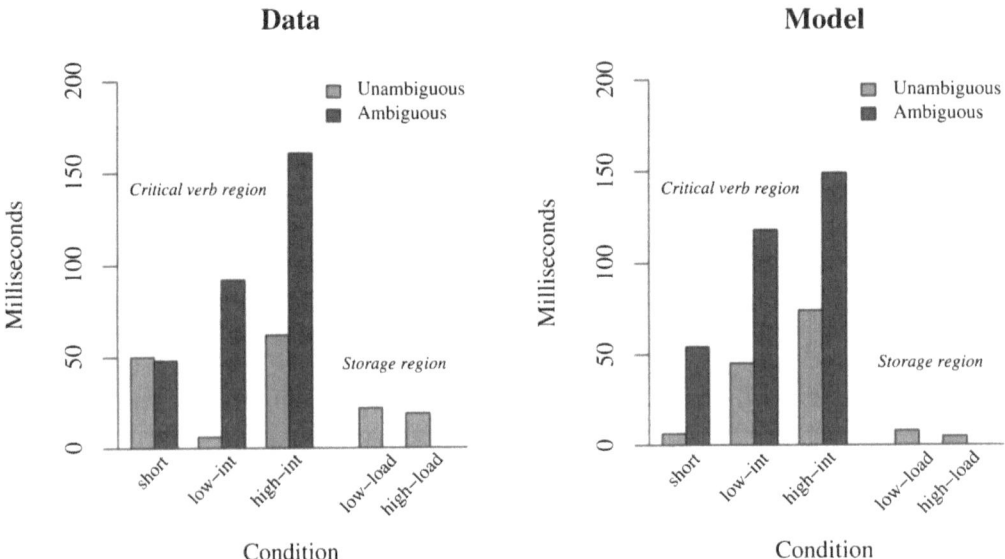

Fig. 8. Model-generated and observed self-paced residual reading times on the critical verb in unambiguous and ambiguous (garden-path) structures as a function of length and structural interference, and reading times in the storage region (Region 2 in the original experiment) as a function of syntactic storage load. Human data from Experiment 4, Van Dyke and Lewis (2003). Short = zero length ambiguous region, low-int(erference) = long ambiguous region with low structural interference, high-int = long ambiguous region with high structural interference. Low-load = low syntactic storage load, high-load = high syntactic storage load.

But we do see a discrepancy—see the first two data points in Fig. 8 and the third part of Table 7. The striking aspect of the data is the complete absence of a length effect on the unambiguous conditions. In fact, the effect numerically goes in the opposite direction: a slight speedup. As a result, the model also underpredicts the effect of length on reanalysis. In short, the actual human data is a more striking reflection of the qualitative form of the prediction than the model behavior is. We believe that the model data overestimates the length effect because it underesti-

Table 7
Contrasts in reading times (msec) at the critical verb for ambiguous and unambiguous structures as a function of length and interference (data from Experiment 4, Van Dyke & Lewis, 2003)

Contrast	Model	Data
Aggregate garden-path effect	65	61
Interference effect		
Aggregate	30	62
On garden-path effect	2	13
On attachment	29	56
Length effect		
Aggregate	52	0
On garden-path effect	25	88
On attachment	39	−44

mates the effect that participating in the parse has on activation level. In particular, there may be additional retrievals associated with semantic processing that are not captured by this model.

5.6. Simulation 5: Effects of storage load

The ACT–R model posits no separate role for "storage" versus "processing" or "integration" costs, in contrast to the Just and Carpenter (1992) models and the storage-based theories of Gibson (1998, 2000). There is only retrieval cost as modulated by associative interference and activation fluctuation. Theoretically, what the ACT–R model claims is that no processing effort is expended in keeping prior constituents active: They simply decay, and the difficulty in reactivating them is a function of how much they have decayed and how effective and discriminating the retrieval cues are. Thus, there is not a general prediction of storage-load effects—instead, there is a more specific prediction of similarity-based retrieval interference.

Grodner, Gibson, and Tunstall (2002) reported a self-paced reading experiment that manipulated syntactic load to test the effects of syntactic storage costs on processing of ambiguous and unambiguous constructions. We focus here on their unambiguous constructions, examples of which are given in (12):

(12) a. [SC/low load] The witness thought that the evidence that was examined by the lawyer implicated his next-door neighbor.
b. [RC/high load] The witness who the evidence that was examined by the lawyer implicated seemed to be very nervous.

The region of interest here is *the evidence that was examined by the lawyer implicated*. In the low-load condition, this region is embedded in a sentential complement. In the high-load condition, this region is embedded in a relative clause, so that a prediction of the empty category must be maintained across the region. Under the storage model, the reading times should increase in this region.

The results, shown in Fig. 9 confirm the predictions of the storage account. We have divided up the regions of interest according to the regions in the original analysis. We collapsed together Regions 3 (*the evidence*) and 4 (*that was*) to take into account spillover effects from the reanalysis of SR to OR clause at the subject NP *the evidence* in (12-b).

The are several important things to note about the model-data comparison. This the first simulation we have presented that models reading times across regions of a sentence that include constructions other than verbs. Such comparisons are inherently problematic because they compare across syntactic and semantic types, but we note that the model is doing a reasonable job of accounting for the basic patterns with syntactic processing alone. This is broadly consistent with the success that Grodner and Gibson (in press) had in modeling reading times with their locality-integration model.

More to the point, however, is whether the model succeeds in capturing the differences between the low- and high-load conditions. Fig. 9 clearly reveals that the success is mixed. In the region *the evidence that was* (Regions 3 and 4 in the original analysis), the model predicts greater reading times in the high-load condition due in part to processing involved in reanalyzing from a subject to OR clause. At the embedded-verb region, the model predicts in-

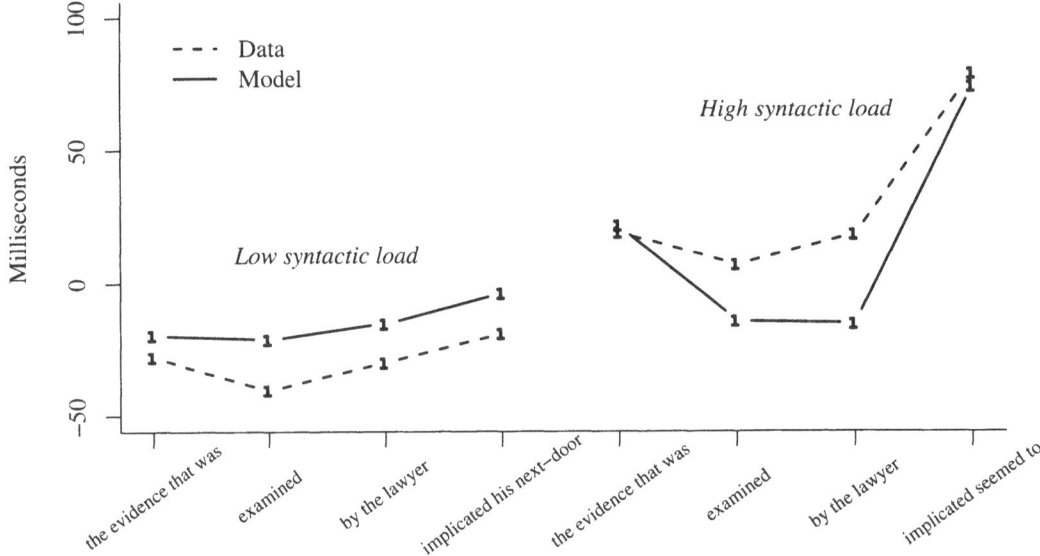

Fig. 9. Model-generated and observed residual self-paced reading times in regions of low and high syntactic storage load. Human data from Experiment 1, Grodner et al. (2002).

creased times associated with filling the object gap in the RC condition, for all the reasons explained earlier.

There is also, perhaps surprisingly, a small load effect at *examined;* this is a retrieval-interference effect due to the additional expected IP. Thus, we see that the model can in principle capture some storage effects as retrieval effects. The model does not presently account for the size of the effect, or the effect on the agent PP *by the lawyer,* although the latter could plausibly be spillover. Table 8 summarizes the contrasts.

One way to distinguish the retrieval interference and storage accounts is to find cases where the locality–storage theory predicts an increase in reading time during the maintenance of a stored prediction, but *before* the critical retrieval. The materials in the previously discussed

Table 8
Contrasts in reading times (msec) as a function of syntactic storage load (data from Grodner et al., 2002, and Van Dyke & Lewis, 2003)

Contrast	Model	Data
Load effect (Grodner et al., 2002)		
Aggregate (not including *implicated*)	48	142
the evidence that was	41	46
examined	7	47
by the lawyer	0	49
implicated...	78	98
Load effect (Van Dyke & Lewis, 2003)		
Storage region	−3	−3

Van Dyke and Lewis (2003) experiment include such a condition. Van Dyke and Lewis detailed how the storage account predicts increased reading time across the italicized regions of the long unambiguous materials (Region 2 in the original analysis):

(13) a. [unambig:long] The assistant forgot that the student who *was waiting for the exam* was standing in the hallway.
      b. [unambig:int] The assistant forgot that the student who *knew that the exam was important* was standing in the hallway.

The reading times revealed no significant load differences, however (see the storage region points in Fig. 8 and the contrast in Table 8). The model actually predicts a tiny decrease in time over the high-load region.

In sum, the model predicts reasonably well the pattern across regions in the two load conditions of Grodner et al. (2002), but only partially accounts for the size of the load effect. However, data from Van Dyke and Lewis (2003) suggests that the explanation of load effects as retrieval-interference effects may in fact be the correct one. A more complete account of these patterns awaits further development of the model, as we describe in the General Discussion section.

6. Center embedding and the problem of serial order discrimination

We turn now to a set of contrasts concerning center embedding. Before we consider the center-embedded structures, it is important to confirm that the model does not predict dramatic increases in reading times for deep right embeddings.

6.1. Deep right embeddings

We ran the model on the following structures: deep right-branching sentential complements, and deep right-branching prepositional phrase modifiers.

(14) a. The boy thought that the man said$_1$ that the editor thought$_2$ that the writer believed$_3$ that the medic thought$_4$ that the assistant admired the student.
      b. The medic with$_1$ the assistant with$_2$ the dog with$_3$ the editor with$_4$ the duck ...

We are interested in the attachment times for the verbs numbered 1 to 4 and the prepositions numbered 1 to 4. The question is whether there is a significant increase as depth of branching goes up. Table 9 gives the results, setting the attachment time of the first SC or PP at zero as a baseline; thus the other times are increases over this baseline. For the embedded sentential complements, there is a tiny increase of about 5 to 6 msec at each depth. This is a small effect of proactive retrieval interference. For the prepositional phrases, there is a larger but still modest increase (of 10 msec, decreasing to 6 msec) at successive attachments. This also represents proactive retrieval interference from additional potential attachment sites. The reason that the PP attachment generates longer increases is informative about the structure of the model: The sentential complements are arguments, and once attached, the verb no longer expects a complement. This keeps the "fan" of the retrieval cue at a constant across the embedding. The re-

Table 9
Memory-retrieval times in the model as a function of right-branching depth

Level	Embedded Sentential Complements (at the verb)	Embedded Noun-Modifying Prepositional Phrases (at the preposition)
1	0	0
2	6	10
3	11	18
4	16	24

Note. The first level is the baseline and set to zero.

trieval cue for the PP modifier attachment, however, is simply a constituent of category NP (or VP), and the number of potential attachment sites continues to increase throughout the sentence.

It seems unlikely that the small increases in the sentential complement embeddings could be detected empirically—even at greater depths, because the size of the effect continues to decrease as a result of the nonlinearity of Equation 3. The PP attachment times might be detectable, and further empirical work is warranted to see if the model's predictions are quantitatively correct. But for present purposes the results of the simulations support the qualitative claim that deep right branching does not produce dramatic effects in processing times.

6.2. The "no serial order representation" hypothesis

As part of a functional analysis of the demands on serial order representation in parsing, Lewis (2000) raised the following possibility: Perhaps there is no serial order representation or serial order mechanism in human sentence processing at all—parsing is based on nothing more than cue-based associative retrievals. This fits well with the findings of McElree (McElree & Dosher, 1989, 1993), which point to rapid, parallel access from working memory of *item* information, but a slow, serial access of relative *order* information.[5]

The ACT–R model embodies this hypothesis. In short, the model proposes that sentence processing relies heavily on discriminating retrieval cues and the distinction between the present and the past. For the most part, this works well—where it potentially fails is in cases that have the following structure, where β is a word that triggers the retrieval of either α_1 or α_2, where α_1 and α_2 cannot be distinguished except on the basis of their relative serial positions:

(15) $\alpha_1 \ldots \alpha_2 \ldots \beta$

In a case such as (15) previously, memory retrieval will succeed for whichever of α_1 or α_2 has the highest activation value. All things being equal, this will be the most recent element due to decay, which is correct for nested structures (but not for cross-serial[6]).

Double self-embeddings are precisely cases that fit this general schema. Consider the extreme case of the classic double-embedded OR clause:

(16) The salmon that the man that the dog bit smoked tasted good.

These structures are notoriously difficult to comprehend (Blauberg & Braine, 1974; Blumenthal, 1966; Hakes & Cairns, 1970; Marks, 1968; Miller & Isard, 1964; Wang, 1970). Several theorists have proposed accounts of the processing difficulty in terms of the demand center embedding places on memory resources of some kind (e.g., Gibson, 1991; Kimball, 1973; Lewis, 1996; Miller & Isard, 1964; Stabler, 1994; Yngve, 1960). We identify here another basic problem with processing such structures: There are multiple attachment points that require distinguishing candidate constituents primarily or exclusively on the basis of their relative serial order. In particular, in (16), there are two active fillers and two predicted embedded finite clauses that must be properly distinguished by serial order to make the correct attachments at the verbs.

6.3. Simulation 6: Toward a graded taxonomy of center embeddings

To test how much various forms of center embedding depend on serial order discrimination, we ran a series of simulations of the model with activation noise.[7] One of ACT–R's standard assumptions is that there is always some noise added to the activation value of a chunk at each retrieval. This noise value is generated from a logistic distribution with mean 0 and variance given by Equation 5:

$$\sigma^2 = \frac{\pi^2}{3} s^2 \tag{5}$$

where s is a parameter. This activation noise permits the use of ACT–R to model RT distributions and various kinds of memory errors; it plays a key role in the Anderson and Matessa (1997) model of list memory, for example, where s ranged from 0.2 to 0.5. We set the noise parameter s to 0.3.

One of the empirical puzzles about double center embedding pointed out by several theorists (Cowper, 1976; Gibson, 1991), and most fully addressed by Gibson (1998), is the existence of double clausal center embeddings that do not yield the same breakdown associated with (16). For example, subject sentences such as (17-a) are noticeably easier that the classic double ORs, as are sentential complements embedded within relatives (17-b; Gibson, 1998):

(17) a. That the food that the medic ordered tasted good pleased her.
     b. The news that the assistant that the lawyer hired left the firm surprised everyone.

We were therefore interested in the model's performance on these structures relative to the classic cases. The full set of structures was as follows:

(18) a. [OR/OR] *object relative within object relative*
         The editor that the writer that the dog chased scolded supervised the assistants.
     b. [SR/OR] *subject relative within object relative*
         The medic who the dog that bit the reporter chased admired the writer.
     c. [RC/SC] *sentential complement within object relative clause*
         The reporter who the claim that the editor admired the medic amused sent the gift.
     d. [SC/SC] *sentential complement within sentential complement*

The claim that the news that the reporter admired the medic amused the editor upset everyone.

e. [SC/RC] *object relative within sentential complement*
   The claim that the reporter who the editor admired sent the gift amused the writer.

f. [SS] *object relative within subject sentence*
   That the reporter who the editor married liked the medic was surprising.

The right-branching structure (14-a) was included as a baseline.

For each structure, we ran the model 100 times and determined the number of times that the model arrived at the correct parse, as well as the number of times that a parse led to a memory-retrieval failure. Not all incorrect parses led to memory-retrieval failures, but in some cases an incorrect attachment earlier in the sentence eliminated the correct candidate for a retrieval later in the sentence. This happened, for example, when the main verb prediction was retrieved too early—perhaps at the second verb in a double center embedding—and was not available when the final (main) verb was actually read.

Fig. 10 plots the results. There are a number of interesting results. The most striking effect is the qualitative contrast between the two difficult double relative clause embeddings and the two easier double embeddings introduced previously.

A second interesting contrast is between the RC/SC and SC/RC embedding: the former led to some misparses. The RC/SC and SC/RC contrast was first pointed out by Gibson (1998); the locality theory predicts the contrast because the filler-gap relation is over a much longer distance in the RC/SC construction than in the SC/RC construction. We visit this contrast again in a moment.

A third interesting and more subtle contrast is between the OR/OR and SR/OR structures. The prediction that the SR/OR structure is easier is consistent with the results of McElree et al.

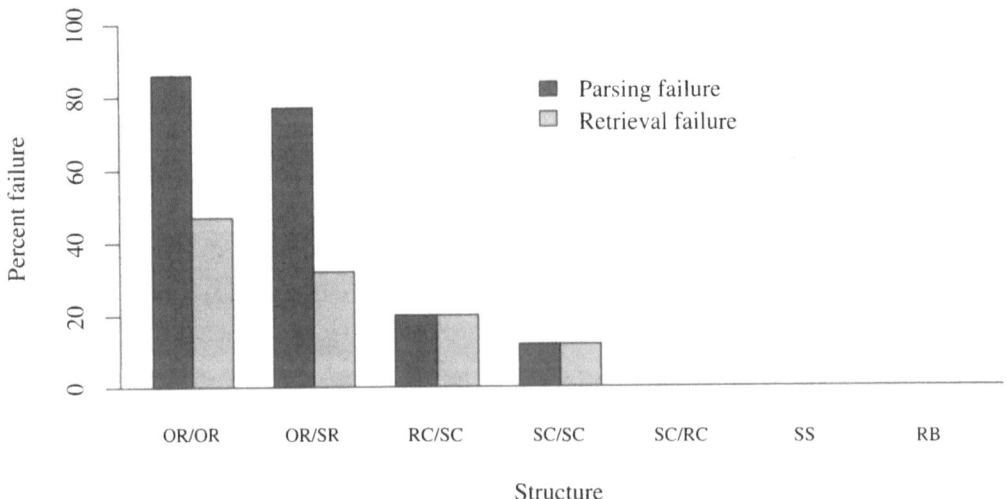

Fig. 10. The behavior of the model on double-center-embeddings, when activation noise is added. Parsing failure is the percentage (out of 100) of over-all failed parses. Retrieval failure is the percentage of parses that led to working-memory retrieval failures.

(2003) who found that the OR/OR structures produced slower dynamics than the SR–OR structures. Their description of how this might have happened in their SAT paradigm is remarkably consistent with our account:

> Misanalysis of any one of the (verb-argument) dependencies will leave a verb stranded without an argument or produce a semantic anomaly. Time course differences of the form seen here could arise from successful reanalysis following misanalysis on a proportion of times. (p. 81)

"Misanalyzing" dependencies is precisely what the model is doing here. A little over half of these misanalyses do result in memory-retrieval failures that result from leaving a verb "stranded." Finally, note the positioning of the SC/SC contrast, somewhere midway between the RC/SC and SC/RC. The increased errors on SC/SC relative to SC/RC are a result of the greater discriminability of the clauses in the SC/RC case—the relative clause carries an additional "gap" feature, which permits a bit more retrieval-time discrimination.

One way to view the ordering of the structures in Fig. 10 is that the failure rates provide a quantitative measure of the degree to which successful parsing of these structures depends on an accurate representation of serial order. This overall ordering is consistent with the acceptability ratings reported by Gibson and Thomas (1997) as well, with the exception that they found that the SC/SC structures patterned with RC/SC—the SC/SC structures were not more difficult.

However, it is important to note that the results of Fig. 10 report parsing failure rates only; they do not take into account reading-time differences. The model's predicted reading times across the three critical verbs for the SC/RC, RC/SC, and SC/SC structures are given in Table 10. Not surprisingly, the highest reading times occur in the SC/RC and RC/SC structures. By far the highest predicted reading time occurs in the RC/SC structure on the second verb. The increased distance of the filler-gap relation means that there is more time for the gap filler's activation to decay—and crucially, there are no intervening reactivations to mitigate this effect. This is consistent with the locality account.

Although we do not yet have a precise quantitative theory that predicts acceptability judgments, we believe a plausible one will take into account *both* failure rates and latencies. The latencies will tend to dominate in structures with low failure rates. Under such an account, it would not be surprising to see the SC/SC structure and SC/RC structures pattern together. The reason is that although the SC/SC structures generated somewhat more failures than SC/RC structures, they also generated somewhat lower reading times when the parses succeeded (see Table 10 and Fig. 10).

Although we believe this is the first explicit serial order discrimination account of difficult center embeddings in the literature, it is very much in the spirit of accounts proposed by a num-

Table 10
Predicted reading times for verbs on SC (sentential complements) and RC (relative clause) mixes

Attachment time	RC/SC	SC/RC	SC/SC
Verb 1	163	278	163
Verb 2	371	210	211
Verb 3	238	237	211

ber of theorists over the years that explain the difficulty in terms of similarity and discrimination rather than memory overload (Bever, 1970; Chomsky, 1965).[7]

7. Summary and general discussion

7.1. Summary: Theory and modeling results

We presented a theory and computational model of the working-memory processes that subserve sentence comprehension. The theory construes parsing as an efficient series of guided memory retrievals. The basic tenets of the theory are simple and follow directly from ACT–R: Sentence processing is controlled by production rules representing the skill of efficient parsing; the activation of memory elements fluctuates as a function of usage and decay; and memory retrieval is subject to associative interference. The computational model represents a novel integration and application of these theoretical concepts to sentence processing.

The model was successful in accounting quantitatively for reading times from five separate experiments, which test in various ways the model's assumptions about memory and processing. The model also led naturally to a novel account of the difficulty of processing center-embedded structures and yielded a graded taxonomy of double center embeddings ranging from easy to difficult.

These results provide support for the claim that the ACT–R account of memory retrieval, developed in domains outside real-time sentence processing, also provides a good account of what is happening in the moment-by-moment retrievals of parsing.

Table 1 summarizes the simulation results and underlying quantitative parameters. Only the single scaling factor was estimated to fit the data, and this factor was set once for all the simulations. All other parameters were set based on ACT–R defaults and prior work on memory interference.

Although the fit to the data is reasonably good for a single-parameter model, it does provide clear cues for how to improve the model: In general, the model appears to *overestimate length effects* and *underestimate interference effects*. Any structural or quantitative change to the model that moves in the direction of decreased emphasis on decay and increased emphasis on interference would likely yield better fits. This bodes well for extending the theory to semantic and discourse processing, which can only add extra retrievals and therefore reduce the effects of decay, but cannot reduce interference.

7.2. Relation to other models and theories

Although the ACT–R-based theory is conceptually simple, it has interesting relations to a wide variety of other approaches in the literature and owes a large intellectual debt to much empirical and modeling work in psycholinguistics and cognitive architectures. We consider some of these debts, spending the most time relating the approach to Gibson's (1998, 2000) DLT, which has been one of the most successful attempts to date to account for data such as that presented here.

7.2.1. Relation to DLT

There are several views one might take on the relation of the ACT–R model to DLT. One view focuses on the fact that the ACT–R model is a decay model and takes the model to simply be an elaborate implementation of locality theory. A second, related, view takes the ACT–R account to be a cognitively plausible, lower level computational realization of locality principles. We believe both views are incorrect. The reason is that DLT and the ACT–R-based accounts diverge substantially, both theoretically and empirically. There are at least four key points of divergence:

1. *Fluctuating activation* is not the same theoretical concept as *locality*. A locality framework leads the theorist to ask the question: What linguistic units should be counted as contributing to locality cost? Indeed, attempting to answer this question has driven some of the work in the DLT framework. ACT–R's activation fluctuation, in contrast, does not require the postulation of a unit of locality cost. The only currency is time. One particularly interesting difference is that DLT posits a difference in integration cost as a function of the referential types of the intervening elements over which integration is made; this is not an explicit part of the ACT–R model.

Furthermore, what counts in the ACT–R model is not the distance to the dependent, but rather the history of retrievals. In fact, ACT–R's activation equation is not properly described as a *decay* equation; rather, it predicts both power law learning and forgetting (Anderson & Lebiere, 1998): The activation can in fact rise over time with additional retrievals.

Of course, we are not denying that there is a clear relation between decay and locality. All things being equal, longer distances (over whatever unit) usually mean more time, which sometimes (but not always) means activation decay. This association of length and decay has a long history in psycholinguistics: It is why Chomsky (1965) suggested that heaviness and length effects point to memory decay as the culprit, and why Gibson (1998) also suggested decay might be behind distance effects. We believe they are both right.

2. *Associative retrieval interference* is not the same theoretical concept as *storage cost*. DLT posits separate integration and storage cost components; this dual processing–storage framework is more consistent with the Just and Carpenter (1992) model than ACT–R. The ACT–R model expends no resource keeping memory items active (of course, it can strategically work to keep items active via rehearsal, but there is not time to do that in language processing). The ACT–R model posits no computational resource consumed except time.

3. *Inadequate serial order discrimination* and *high locality-based integration costs* are different theoretical explanations for the difficulty associated with the notoriously bad double center embeddings. As we outlined previously, the discrimination account does not supplant the distance and interference effects, which the model also captures in the reading times on those rare occasions that it parses them correctly. But we believe that the discrimination account provides a plausible explanation for the rather dramatic qualitative effect of the double embeddings, and complements in an important way a reading-time–locality-based explanation.

4. The ACT–R model specifies computational structure and quantitative parameters on an independent basis that must be added to DLT to generate precise quantitative predictions. In particular, these include assumptions about the locality cost function (linear, nonlinear, etc.) and the combination function for cost at points of multiple integrations (e.g., whether the dis-

tances are summed then put through the cost function, or put through the cost function separately and then summed, etc.)

The empirical predictions of DLT and the ACT–R model reflect the theoretical divergence; we noted some of these throughout the article. To summarize:

1. The fluctuating activation of the ACT–R model (which can yield different levels of activation across identical distances) was critical for accounting for the differential effect of distance and interference on garden-path reanalysis and attachment (Simulation 4). (2) The ACT–R model makes different predictions from DLT about "storage effects" (Simulations 4 and 5). (3) The ACT–R model predicts a cross-over interaction of extraction type and verb (main vs. embedded) in the reading times on relative clauses (Simulation 1); DLT does not. (4) The ACT–R model predicts a greater effect of interpolated material on embedded relative clauses than main verbs (due to greater interference and processing of the object-gap relation); DLT predicts this interaction but to a smaller degree than the ACT–R model (Simulation 3). (5) The ACT–R model predicts a relatively flat effect of interpolated material on main verb reading times (Simulations 1–3); this is a direct result of the parallel associative retrieval mechanisms of the model. (6) The ACT–R model predicts probabilistic parsing failures in addition to longer reading times; DLT has not been used to generate such probabilistic predictions (Simulation 6; see Gibson & Thomas, 1999, for use of locality theory to account for binary acceptability contrasts). (7) Although we did not show such effects in the simulations in this article, the ACT–R model is capable of predicting "reverse" length effects—that is, *faster* reading times with increased length. Such apparently puzzling effects have been confirmed now in German (Konieczny, 2000; Vasishth, Cramer, Scheepers, & Wagner, 2003; Vasishth, Scheepers, Uszkoreit, Wagner, & Kruijff, 2004), Hindi (Vasishth, 2003a, 2003b), and Japanese (Gibson, personal communication, July 2003). We are currently developing models to account for data from these languages.

Finally, although we have not yet attempted to model differences due to referential type of intervening constituents, we expect differences in the predictions of the ACT–R model and DLT, which explicitly assumes that integration cost varies as a function of referential type. One important challenge for the ACT–R model is to account for the greater ease of processing double center embeddings with pronouns in the subjects of the most embedded clause (Bever, 1970; Gibson, 1998; Gordon et al., 2001; Kac, 1981). ACT–R would clearly be sensitive to the difference because the pronoun is processed more quickly than a full NP, but it seems unlikely that this would be enough to capture the contrast in the difficult embeddings. However, one possibility is to simply assume that nominals have base activations that reflect their referential type (perhaps their position in an accessibility hierarchy; see Givón, 1978). Such an assumption does not derive from the ACT–R theory, itself, unlike other aspects of the theory presented here—it would simply be an assumption about how certain linguistic distinctions are manifest in processing. This is precisely the status of the assumption in DLT.

7.2.2. Working-memory theory

As we pointed out earlier, the ACT–R theory is most consistent with the memory theory of McElree (1998), which emphasizes an extremely limited working-memory focus and parallel, associative retrieval. What this theory adds is activation fluctuation and similarity-based retrieval interference. One minor, but possibly empirically very important, structural difference is that the McElree account assumes that the most recent item is available in the focus. That is not always possible in the ACT–R model, depending on how that item was processed. In general, the McElree SAT phenomena themselves will provide important future empirical tests of the model.

7.2.2.1. Activation-based models. Of the prior computational sentence-processing theories, the ACT–R model bears the most resemblance to other hybrid symbolic activation-based models, including the competition model of Stevenson (1994), the inhibition model of Vosse and Kempen (2000), and CAPS/CC-Reader (Just & Carpenter, 1992). The Stevenson model has something like an activation decay component and predicts recency effects in a similar way to ACT–R. It maintains a stack-like data structure, however, so we do not believe it will provide the foundation for a plausible account of unambiguous structures. The Vosse and Kempen (2000) unification space model has been applied to both ambiguous and unambiguous structures, and its account of difficult center embeddings is the closest to the one we proposed here. It has been used to model qualitative garden-path and embedding phenomena and has neurobiological support (Hagoort, 2003). However, it was not intended to—and is unable to (Kempen, personal communication, September 2004)—model reading-time data. Although all of these models use the concept of activation of memory elements, the ACT–R account has the virtue that it is founded on a set of independently motivated and extensively tested architectural principles of memory retrieval and controlled processing; the success of the reported modeling simulations attests to the benefits of this foundation.

7.2.2.2. Interference sentence-processing theories. Some recent approaches to sentence processing have emphasized the role of similarity-based interference (Gordon et al., 2001; Gordon, Hendrick, & Levine, 2002; Lewis, 1993, 1996; Van Dyke & Lewis, 2003). Lewis (1993, 1996) emphasized *storage* interference, whereas Lewis (1998b) and Gordon et al. (2001, 2002) emphasized *retrieval* interference. The latter emphasis is consistent with our current account.

It is worth reviewing briefly here why similarity-interference arises in the model, and how similarity is computed. Similarity interference arises at retrieval, and it is a direct consequence of Equation 3, which reduces the strength of association between a cue and a target as a function of the number of items associated with that cue. Reduced strength of association means reduced activation boost, which produces higher latencies and higher failure rates. The strong claim about this kind of interference is that it is *retrieval* interference, and it is purely a function of *cue overlap*. In this model, the retrieval cues are simply a subset of the features of a target item, so similarity interference is based purely on representational overlap, not general thematic connections (e.g., *cow–milk,* though ACT–R as an architecture admits of such associations).

It is therefore the nature of the cues that dictate the nature of the interference. In this model, we have realized only syntactic cues, which are used primarily to reactivate predicted structure to unify with. However, the model can accommodate a richer set of cues—for example, there may also be semantic cues derived from specific lexical constraints (e.g., the semantic constraints that a verb places on its subject). Apart from the extensive literature on the immediate effects of semantic constraints on ambiguity resolution (e.g., Trueswell et al., 1994), Van Dyke (2002) provided direct evidence for interference effects as a function of semantic cue overlap. Adopting such a richer set of cues may be on the path to providing an account for Gordon's (Gordon et al., 2001; Gordon, Hendrick, & Johnson, 2004) effects of NP type—at least on reading times at the verb, when retrieval interference should arise. However, some of the similarity effects—particularly those showing up before the verb (Gordon et al., 2004)—may be more properly understood as *encoding* interference, something that is missing from all current processing models.

7.2.3. Degrees of freedom in the modeling

A legitimate concern that can be raised about applying computational cognitive architectures to empirical data is whether the theorist has too many degrees of freedom in fitting data. We will not attempt to provide a general response to this concern (for insightful responses, see Anderson et al., 2005; Meyer & Kieras, 1997; Newell, 1990). We will attempt to address the issue with respect to this model.

As with any ACT–R model, there are two kinds of degrees of freedom: quantitative parameters, and the contents of production rules and declarative memory. The quantitative parameters, such as the decay rate, associative strength, and so on, are the easiest to map onto the traditional concept of degrees of freedom in statistical modeling. With respect to these parameters, our response is straightforward: We believe we have come as close as possible to zero-parameter predictions by adjusting only the scaling factor, adopting all other values from the ACT–R defaults and literature, and using the same parameter settings across multiple simulations.

With respect to the degrees of freedom inherent in the content of the production rules and declarative memory, we take advantage of the fact that in sentence processing we are blessed with a brutally fast domain with a well-studied content domain theory (syntax). The absolute time scale of the effects means that there is simply no room for adding or subtracting a production here and there to get a better fit. The overarching principle for language processing is "Do things as fast as possible," and that principle entirely guided the development of the parser. All of the model's processing is done in the most efficient manner that is functionally possible, and all of it boils down to two or three production firings and one or two memory retrievals.

7.2.4. The value of architectures and prospects for the future

Is ACT–R just an implementation language for theories? Again, we will not provide a general answer to the question, but simply note that the answer should be very clear in this case. ACT–R is not simply an implementation language for the theory; a comparison of Tables 2 and 3 reveals that this sentence-processing theory is itself best understood as the application of ACT–R theory (and thus, a particular set of theoretical principles) to the task of parsing.

There are many interesting possible directions for additional functional and empirical extensions to the model. It should be possible to extend the model to lower levels of processing to explore the interactions of lexical and syntactic processing; indeed much of the structure is already in place to do so. It should also be possible to extend the model to higher levels of semantic and discourse processing; an integration with the semantic interpretation model of Budiu and Anderson (2004) provides one possible avenue. Given prior ACT–R work on modeling individual differences, it should also be possible to use this model as a basis for accounting for performance differences due to variation in working-memory capacity.

Another important direction is to develop detailed models of the different experimental paradigms used in reading research, including eye-tracking, self-paced reading, central presentation, and speed–accuracy–trade-off designs. Developing such models requires making explicit linking assumptions about how linguistic processing is coordinated with eye-movement control and manual button pressing. It is one of the key advantages of working within an architecture such as ACT–R that such models can be developed in a theoretically constrained and computationally explicit manner. This also has the advantage of addressing a possible concern about relying too heavily on the self-paced moving-window paradigm, which might increase working-memory demands relative to natural reading.

Just as important, we believe the model opens up new empirical and theoretical questions to ask about sentence processing: questions about the precise nature of length effects, interference effects, and the relation of linguistic processing to more general cognitive processing. These are all direct benefits of pursuing the architectural approach to cognitive theory pioneered by Newell (Newell, 1990; Newell & Simon, 1972) and Anderson (Anderson, 1983, Anderson, 2005, this issue).

Notes

1. We make a simplifying assumption in the present model that adjuncts are attached via a "modifier" slot rather than through Chomsky-adjunction, though this is not a critical assumption for the model's empirical predictions.

2. This presentation borrows from Crocker (1996), and the rules depart from our X-bar assumptions in some ways.

3. The distance-based explanation does not appear to work for head-final languages because the verb invariably appears clause finally in the embedded clause, and yet in these languages too there is an SO preference. Here, Korthals' frequency-based explanation is plausible.

4. One might argue that there should be additional retrievals for VP-internal subject traces; doing so would just increase times for all the verbs across the board and would not change the basic pattern.

5. McElree et al. (2003) adopt an alternative position: that the sentence processing has special-purpose serial order mechanisms that operate on a fast, parallel basis. We believe that the data that motivated this position can be handled by the model without serial information; but this remains a tenable position.

6. There are a number of linguistic phenomena that may appear at first to be insurmountable challenges to parsing without serial order. These include explicit order terms such as *former*, *latter*, and *respectively*, and cross-serial dependencies such as those that routinely appear in Dutch subordinate clauses with perception or causation verbs (Bach, Brown, & Marslen-Wilson, 1986; Kaan & Vasic, 2004; Steedman, 2000). Cross-serial dependencies are particularly interesting because they are syntactic constructions that appear to violate the standard nested most-recent-first ordering most naturally sup-

ported by a parser with activation decay. A full treatment of these phenomena is beyond the scope of this paper, but it is worth summarizing briefly what our initial analyses have revealed. First, it is important to carefully analyze what other cues may be available to the system to achieve the discrimination required—even in the absence of serial order information. Perhaps surprisingly, in a LC parse of Dutch cross-serial dependencies, there is enough information available at the verb to discriminate the predicted verbal categories *without* appeal to explicit serial order information, and without additional processing cost relative to nested structures. Furthermore, even in cases where explicit serial order information does seem to be required (as we believe is the case for order terms), the locus of that order information may be in a different memory representation from that supporting the incremental parse. Our hypothesis is that order terms like *former* and *latter*, or *first*, *second* and *third*, are discourse anaphors whose semantics are grounded in explicit relations in a discourse model, perhaps held in long-term working memory. In this sense, they may not be fundamentally different from spatial relation terms like *above* or *below* (and indeed such terms are often used in text discourse, as we have occasionally done above).

7. The other simulations presented in this paper do not have activation noise switched on because the goal there was to first understand the model's behavior without involving any random variation. We have started to experiment with activation noise in these other simulations, and preliminary results suggest that the results do not change significantly.

Acknowledgments

We thank Daniel Grodner for generously sharing a prepublication version of his work with Ted Gibson. We are also grateful to Jerry Ball and three anonymous reviewers whose detailed comments significantly improved this article. Please send comments to either of us at rickl@umich.edu or vasishth@acm.org

References

Aho, A. V., & Ullman, J. D. (1972). *The theory of parsing, translation, and compiling: Vol.1. Parsing.* Englewood Cliffs, NJ: Prentice Hall.

Altmann, G., & Steedman, M. (1988). Interaction with context during human sentence processing. *Cognition, 30,* 191–238.

Anderson, J. R. (1983). *The architecture of cognition.* Cambridge, MA: Harvard University Press.

Anderson, J. R. (2005). Human symbol manipulation within an integrated cognitive architecture. *Cognitive Science, 29,* 313–341.

Anderson, J. R., Bothell, D., Byrne, M. D., Douglass, S., Lebiere, C., & Qin, Y. (2005). An integrated theory of mind. *Psychological Review.*

Anderson, J. R., & Lebiere, C. (1998). *Atomic components of thought.* Hillsdale, NJ: Lawrence Erlbaum Associates, Inc.

Anderson, J. R., & Matessa, M. P. (1997). A production system theory of serial memory. *Psychological Review, 104,* 728–748.

Anderson, J. R., & Reder, L. M. (1999). The fan effect: New results and new theories. *Journal of Experimental Psychology: General, 128,* 186–197.

Anderson, J. R., & Schooler, L. J. (1991). Reflections of the environment in memory. *Psychological Science, 2,* 396–408.

Bach, E., Brown, C., & Marslen-Wilson, W. (1986). Crossed and nested dependencies in German and Dutch: A psycholinguistic study. *Language and Cognitive Processes, 1,* 249–262.

Bader, M., & Lasser, I. (1994). German verb-final clauses and sentence processing: Evidence for immediate attachment. In C. Clifton Jr., L. Frazier, & K. Rayner (Eds.), *Perspectives in sentence processing* (pp. 225–242). Hillsdale, NJ: Lawrence Erlbaum Associates, Inc.

Bar-Hillel. (1953). A quasi-arithmetical notation for syntactic description. *Language, 29,* 47–58.

Bever, T. G. (1970). The cognitive basis for linguistic structures. In J. R. Hayes (Ed.), *Cognition and the development of language* (pp. 279–362). New York: Wiley.

Blauberg, M. S., & Braine, M. D. S. (1974). Short-term memory limitations on decoding self-embedded sentences. *Journal of Experimental Psychology, 102,* 745–748.

Blumenthal, A. L. (1966). Observations with self-embedded sentences. *Psychonomic Science, 6,* 453–454.

Bock, J. K. (1986). Syntactic persistence in language production. *Cognitive Psychology, 18,* 355–387.

Bock, J. K., & Griffin, Z. M. (2000). The persistence of structural priming: Transient activation or implicit learning? *Journal of Experimental Psychology: General, 129,* 177–192.

Boland, J. E., & Blodgett, A. (2001). Understanding the constraints on syntactic generation: Lexical bias and discourse congruency effects on eye movements. *Journal of Memory and Language, 45,* 391–411.

Boland, J. E., Lewis, R. L., & Blodgett, A. (2004). *Distinguishing generation and selection of modifier attachments: Implications for lexicalized parsing.* Manuscript submitted for publication.

Budiu, R., & Anderson, J. R. (2004). Interpretation-based processing: A unified theory of semantic sentence comprehension. *Cognitive Science, 28,* 1–44.

Chomsky, N. (1965). *Aspects of the theory of syntax.* Cambridge, MA: MIT Press.

Chomsky, N. (1980). Rules and representations. *Behavioral and Brain Sciences, 3*(1), 1–15.

Chomsky, N. (1986). *Barriers.* Cambridge, MA: MIT Press.

Chomsky, N., & Miller, G. A. (1963). Introduction to the formal analysis of natural languages. In D. R. Luce, R. R. Bush, & E. Galanter (Eds.), *Handbook of mathematical psychology* (Vol. 2; pp. 269–322). New York: John Wiley.

Clifton, C., Jr., & Duffy, S. A. (2001). Sentence and text comprehension: Roles of linguistic structure. *Annual Review of Psychology, 52,* 167–196.

Cowan, N. (2001). The magical number 4 in short-term memory: A reconsideration of mental storage capacity. *Brain and Behavioral Sciences, 24,* 87–114.

Cowper, E. A. (1976). *Constraints on sentence complexity: A model for syntactic processing.* Unpublished doctoral dissertation, Brown University, Providence, RI.

Crocker, M. W. (1996). *Mechanisms for sentence processing* (Research paper EUCCS/RP-70). Edinburgh, Scotland: University of Edinburgh, Centre for Cognitive Science.

Crocker, M. W., & Brants, T. (2000). Wide-coverage probabilistic sentence processing. *Journal of Psycholinguistic Research, 29,* 647–669.

Dell, G. S., Burger, L. K., & Svec, W. R. (1997). Language production and serial order: A functional analysis and a model. *Psychological Review, 104,* 123–147.

Elman, J. L. (1990). Finding structure in time. *Cognitive Science, 14,* 179–211.

Ferreira, F., & Clifton, C., Jr. (1986). The independence of syntactic processing. *Journal of Memory and Language, 25,* 348–368.

Ferreira, F., & Henderson, J. M. (1991). Recovery from misanalyses of garden-path sentences. *Journal of Memory and Language, 30,* 725–745.

Fodor, J. A. (1983). *Modularity of mind: An essay on faculty psychology.* Cambridge, MA: MIT Press.

Ford, M. (1983). A method for obtaining measures of local parsing complexity throughout sentences. *Journal of Verbal Learning and Verbal Behavior, 22,* 203–218.

Frazier, L. (1987a). Sentence processing: A tutorial review. In M. Coltheart (Ed.), *Attention and performance xii* (pp. 559–585). Norwood, NJ: Lawrence Erlbaum Associates, Inc.

Frazier, L. (1987b). Syntactic processing: Evidence from Dutch. *Natural Language and Linguistic Theory, 5,* 519–560.

Frazier, L. (1989). Filler driven parsing: A study of gap filling in Dutch. *Journal of Memory and Language, 28,* 331–334.

Frazier, L. (1995). Constraint satisfaction as a theory of sentence processing. *Journal of Psycholinguistic Research, 6,* 437–468.

Frazier, L. (1998). Getting there … slowly. *Journal of Psycholinguistic Research, 27,* 123–146.

Frazier, L., & Rayner, K. (1982). Making and correcting errors during sentence comprehension: Eye movements in the analysis of structurally ambiguous sentences. *Cognitive Psychology, 14,* 178–210.

Gazdar, G. (1981). Unbounded dependencies and coordinate structure. *Linguistic Inquiry, 12,* 155–184.

Gazdar, G., Klein, E., Pullum, G. K., & Sag, I. A. (1985). *Generalized phrase structure grammar.* Oxford: Basil Blackwell.

Gibson, E. A. (1991). *A computational theory of human linguistic processing: Memory limitations and processing breakdown* (Tech. Rep. CMU-CMT-91–125). Pittsburgh, PA: Carnegie Mellon University, Center for Machine Translation.

Gibson, E. A. (1998). Linguistic complexity: Locality of syntactic dependencies. *Cognition, 68,* 1–76.

Gibson, E. A. (2000). The dependency locality theory: A distance-based theory of linguistic complexity. In Y. Miyashita, A. Mirantz, & W. O'Neil (Eds.), *Image, language, brain* (pp. 95–126). Cambridge, MA: MIT Press.

Gibson, E. A., & Thomas, J. (1997). *The complexity of nested structures in English: Evidence for the syntactic prediction locality theory of linguistic complexity.* Unpublished manuscript, Massachusetts Institute of Technology.

Gibson, E. A., & Thomas, J. (1999). Memory limitations and structural forgetting: The perception of complex ungrammatical sentences as grammatical. *Language and Cognitive Processes, 14,* 225–248.

Givón, T. (1978). Definiteness and referentiality. In J. H. Greenberg, C. A. Ferguson, & E. A. Moravcsik (Eds.), *Universals of human language: Vol. 4. Syntax* (pp. 291–330). Stanford, CA: Stanford University Press.

Gordon, P. C., Hendrick, R., & Johnson, M. (2001). Memory interference during language processing. *Journal of Experimental Psychology: Learning, Memory, and Cognition, 27,* 1411–1423.

Gordon, P. C., Hendrick, R., & Johnson, M. (2004). Effects of noun phrase type on sentence complexity. *Journal of Memory and Language, 51,* 97–114.

Gordon, P. C., Hendrick, R., & Levine, W. H. (2002). Memory load interference in syntactic processing. *Psychological Science, 13,* 425–430.

Green, G. M. (1976). Main clause phenomena in subordinate clauses. *Language, 52,* 382–397.

Grodner, D., Gibson, E., & Tunstall, S. (2002). Syntactic complexity in ambiguity resolution. *Journal of Memory and Language, 46,* 267–295.

Grodner, D. J., & Gibson, E. A. (in press). Consequences of the serial nature of linguistic input for sentential complexity. *Cognitive Science.*

Hagoort, P. (2003). How the brain solves the binding problem for language: A neurocomputational model of syntactic processing. *NeuroImage, 20,* S18–S29.

Hakes, D. T., & Cairns, H. S. (1970). Sentence comprehension and relative pronouns. *Perception & Psychophysics, 8*(1), 5–8.

Hakes, D. T., Evans, J. S., & Brannon, L. L. (1976). Understanding sentences with relative clauses. *Memory and Cognition, 4,* 283–290.

Hakuta, K. (1981). Grammatical description versus configurational arrangement in language acquisition: The case of relative clauses in Japanese. *Cognition, 9,* 197–236.

Hirose, Y., & Inoue, A. (1998). Ambiguities in parsing complex sentences in Japanese. In D. Hillert (Ed.), *Syntax and semantics* (Vol. 31, pp. 71–93). New York: Academic.

Holmes, V. M., & O'Regan, J. K. (1981). Eye fixation patterns during the reading of relative clause sentences. *Journal of Verbal Learning and Verbal Behavior, 20,* 417–430.

Hooper, J. B., & Thompson, S. A. (1973). On the applicability of root transformations. *Linguistic Inquiry, 4,* 465–497.

Hudgins, J. C., & Cullinan, W. L. (1978). Effects of sentence structure on sentence elicited imitation responses. *Journal of Speech and Hearing Research, 21,* 809–819.

Inoue, A., & Fodor, J. D. (1995). Information-paced parsing of Japanese. In R. Mazuka & N. Nagai (Eds.), *Japanese sentence processing* (pp. 9–63). Mahwah, NJ: Lawrence Erlbaum Associates, Inc.

Johnson-Laird, P. N. (1983). *Mental models*. Cambridge, MA: Harvard University Press.

Jurafsky, D. (1996). A probabilistic model of lexical and syntactic access and disambiguation. *Cognitive Science, 20*, 137–194.

Just, M. A., & Carpenter, P. A. (1992). A capacity theory of comprehension: Individual differences in working memory. *Psychological Review, 99*, 122–149.

Kaan, E., & Vasic, N. (2004). Cross-serial dependencies in Dutch: Testing the influence of NP type on processing load. *Memory & Cognition, 32*, 175–184.

Kimball, J. (1973). Seven principles of surface structure parsing in natural language. *Cognition, 2*, 15–47.

King, J., & Just, M. A. (1991). Individual differences in syntactic processing: The role of working memory. *Journal of Memory and Language, 30*, 580–602.

Konieczny, L. (2000). Locality and parsing complexity. *Journal of Psycholinguistic Research, 29*, 627–645.

Konieczny, L., Hemforth, B., Scheepers, C., & Strube, G. (1997). The role of lexical heads in parsing: Evidence from German. *Language and Cognitive Processes, 12*, 307–348.

Korthals, C. (2001). Self embedded relative clauses in a corpus of German newspaper texts. In K. Striegnitz (Ed.), *Proceedings of the Sixth ESSLLI Student Session* (pp. 179–190). Helsinki, Finland: University of Helsinki.

Koster, J. (1978). Why subject sentences don't exist. In S. J. Keyser (Ed.), *Recent transformational studies in European languages* (pp. 53–64). Cambridge, MA: MIT Press.

Larkin, W., & Burns, D. (1977). Sentence comprehension and memory for embedded structure. *Memory and Cognition, 5*, 17–22.

Lewis, R. L. (1993). *An architecturally-based theory of human sentence comprehension* (Tech. Rep. CMU-CS-93-226). Retrieved March 18, 2005, from http://www-personal.umich.edu/~rickl/Documents/thesis.pdf

Lewis, R. L. (1996). Interference in short-term memory: The magical number two (or three) in sentence processing. *Journal of Psycholinguistic Research, 25*, 93–115.

Lewis, R. L. (1998a). Leaping off the garden path: Reanalysis and limited repair parsing. In J. Fodor & F. Ferreira (Eds.), *Reanalysis in sentence processing* (pp. 247–285). Boston: Kluwer Academic.

Lewis, R. L. (1998b, March). *Working memory in sentence processing: Retroactive and proactive interference in parsing*. Paper presented at The 11th Annual CUNY Conference on Human Sentence Processing, New Brunswick, NJ.

Lewis, R. L. (2000). Specifying architectures for language processing: Process, control, and memory in parsing and interpretation. In M. W. Crocker, M. Pickering, & C. Clifton Jr. (Eds.), *Architectures and mechanisms for language processing* (pp. 56–89). Cambridge, England: Cambridge University Press.

Lovett, M. (1998). Choice. In J. R. Anderson & C. Lebiere (Eds.), *The atomic components of thought* (pp. 255–296). Hillsdale, NJ: Lawrence Erlbaum Associates, Inc.

MacDonald, M. C., & Christiansen, M. (2002). Reassessing working memory: Comment on Just and Carpenter (1992) and Waters and Caplan (1996). *Psychological Review, 109*, 35–54.

MacDonald, M. C., Pearlmutter, N. J., & Seidenberg, M. S. (1994). The lexical nature of syntactic ambiguity resolution. *Psychological Review, 101*, 676–703.

Marks, L. E. (1968). Scaling of grammaticalness of self-embedded English sentences. *Journal of Verbal Learning and Verbal Behavior, 7*, 965–967.

McElree, B. (1993). The locus of lexical preference effects in sentence comprehension: A time-course analysis. *Journal of Memory and Language, 32*, 536–571.

McElree, B. (1998). Attended and non-attended states in working memory: Accessing categorized structures. *Journal of Memory and Language, 38*, 225–252.

McElree, B., & Dosher, B. A. (1989). Serial position and set size in short-term memory: Time course of recognition. *Journal of Experimental Psychology: General, 118*, 346–373.

McElree, B., & Dosher, B. A. (1993). Serial retrieval processes in the recovery of order information. *Journal of Experimental Psychology: General, 122*, 291–315.

McElree, B., Foraker, S., & Dyer, L. (2003). Memory structures that subserve sentence comprehension. *Journal of Memory and Language, 48*, 67–91.

McKoon, G., & Ratcliff, R. (2003). Meaning through syntax: Language comprehension and the reduced relative clause construction. *Psychological Review, 110,* 490–525.

Meyer, D. E., & Kieras, D. E. (1997). A computational theory of executive cognitive processes and multiple-task performance: Part 1. Basic mechanisms. *Psychological Review, 104,* 3–65.

Miller, G. A. (1956). The magical number seven plus or minus two: Some limits on our capacity for processing information. *Psychological Review, 63,* 81–97.

Miller, G. A., & Chomsky, N. (1963). Finitary models of language users. In D. R. Luce, R. R. Bush, & E. Galanter (Eds.), *Handbook of mathematical psychology* (Vol. 2, pp. 419–492). New York: Wiley.

Miller, G. A., & Isard, S. (1964). Free recall of self-embedded English sentences. *Information and Control, 7,* 292–303.

Mitchell, D. C. (1994). Sentence parsing. In M. A. Gernsbacher (Ed.), *Handbook of psycholinguistics* (pp. 375–409). San Diego: Academic.

Newell, A. (1973). Production systems: Models of control structures. In W. G. Chase (Ed.), *Visual information processing* (pp. 463–526). New York: Academic Press.

Newell, A. (1990). *Unified theories of cognition.* Cambridge, MA: Harvard University Press.

Newell, A., & Simon, H. A. (1972). *Human problem solving.* Englewood Cliffs, NJ: Prentice Hall.

Pollard, C., & Sag, I. (1994). *Head-driven phrase structure grammar.* Chicago: University of Chicago Press.

Schabes, Y., & Joshi, A. K. (1991). Parsing in lexicalized tree adjoining grammar. In M. Tomita (Ed.), *Current issues in parsing technologies* (pp. 25–47). Boston: Kluwer Academic.

Schlesinger, I. M. (1975). Why a sentence in which a sentence in which a sentence is embedded is embedded is difficult. *International Journal of Psycholinguistics,* 53–66.

Sheldon, A. (1974). The role of parallel function in the acquisition of relative clauses in English. *Journal of Verbal Learning and Verbal Behavior, 13,* 272–281.

Simon, H. A. (1974, February 8). How big is a chunk? *Science, 183,* 482–488.

Srinivas, B., & Joshi, A. K. (1999). Supertagging: An approach to almost parsing. *Computational Linguistics, 25,* 237–265.

Stabler, E. P. (1994). The finite connectivity of linguistic structure. In C. Clifton Jr., L. Frazier, & K. Rayner (Eds.), *Perspectives in sentence processing* (pp. 303–336). Hillsdale, NJ: Lawrence Erlbaum Associates, Inc. (An earlier version of this paper appeared in UCLA Occasional Papers No.11.)

Steedman, M. (1996). *Surface structure and interpretation.* Cambridge, MA: MIT Press.

Steedman, M. (2000). *The Syntactic Process.* Cambridge, MA: MIT Press.

Stevenson, S. (1994). Competition and recency in a hybrid network model of syntactic disambiguation. *Journal of Psycholinguistic Research, 23,* 295–322.

Tabor, W., Juliano, C., & Tanenhaus, M. K. (1998). Parsing in a dynamical system: An attractor-based account of the interaction of lexical and structural constraints in sentence processing. *Language and Cognitive Processes, 12,* 211–272.

Tanenhaus, M. K., Spivey-Knowlton, M. J., Eberhard, K. M., & Sedivy, J. C. (1995). Integration of visual and linguistic information in spoken language comprehension. *Science, 268,* 1632–1634.

Trueswell, J. C., Tanenhaus, M. K., & Garnsey, S. M. (1994). Semantic influences in parsing: Use of thematic role information in syntactic ambiguity resolution. *Journal of Memory and Language, 33,* 285–318.

Ullman, M. T. (2004). Contributions of memory circuits to language: The declarative/procedural model. *Cognition, 92,* 231–270.

Van Dyke, J. A. (2002). *Retrieval effects in sentence parsing and interpretation.* Unpublished doctoral dissertation, University of Pittsburgh, Pittsburgh, PA.

Van Dyke, J. A., & Lewis, R. L. (2003). Distinguishing effects of structure and decay on attachment and repair: A cue-based parsing account of recovery from misanalyzed ambiguities. *Journal of Memory and Language, 49,* 285–316.

van Noord, G. (1997). An efficient implementation of the head-corner parser. *Computational Linguistics, 23,* 425–456.

Vannest, J., Newman, A., Polk, T. A., Lewis, R. L., Newport, E. L., & Bavelier, D. (2004). *fMRI evidence that some, but not all, derived words are decomposed during lexical access.* Manuscript submitted for publication.

Vannest, J., Polk, T. A., & Lewis, R. L. (2003, in revision). Dual-route processing of complex words: New fMRI evidence from derivational suffixation. *Cognitive, Affective, and Behavioral Neuroscience.*

Vasishth, S. (2003a). Quantifying processing difficulty in human sentence parsing: The role of decay, activation, and similarity-based interference. In *Proceedings of the European Cognitive Science Conference (Osnabrück).* Mahwah, NJ: Lawrence Erlbaum Associates, Inc.

Vasishth, S. (2003b). *Working memory in sentence comprehension: Processing Hindi center embeddings.* New York: Garland.

Vasishth, S., Cramer, I., Scheepers, C., & Wagner, J. (2003). Does increasing distance facilitate processing? In *Proceedings of the 16th Annual CUNY Sentence Processing Conference.* MIT, Cambridge, MA (Available from: ftp://ftp.coli.uni-sb.de/pub/people/vasishth/ wip.html)

Vasishth, S., Scheepers, C., Uszkoreit, H., Wagner, J., & Kruijff, G. J. M. (2004, March). *Constraint defeasibility and concurrent constraint satisfaction in human sentence processing.* Paper presented at the CUNY Sentence Processing Conference, Cambridge, MA.

Vosse, T., & Kempen, G. (2000). Syntactic structure assembly in human parsing: A computational model based on competitive inhibition and a lexicalist grammar. *Cognition, 75,* 105–143.

Wang, M. D. (1970). The role of syntactic complexity as a determiner of comprehensibility. *Journal of Verbal Learning and Verbal Behavior, 9,* 398–404.

Wickelgren, W. A., Corbett, A. T., & Dosher, B. A. (1980). Priming and retrieval from short-term memory: A speed-accuracy trade-off analysis. *Journal of Verbal Learning and Verbal Behavior, 19,* 387–404.

Yngve, V. H. (1960). A model and an hypothesis for language structure. *Proceedings of the American Philosophical Society, 104,* 444–466.

Cognitive Science 29 (2005) 421–455

Modeling Parallelization and Flexibility Improvements in Skill Acquisition: From Dual Tasks to Complex Dynamic Skills

Niels Taatgen

Department of Psychology, Carnegie Mellon University
and
Department of Artificial Intelligence, University of Groningen, Netherlands

Received 30 April 2004; received in revised form 19 November 2004; accepted 2 February 2005

Abstract

Emerging parallel processing and increased flexibility during the acquisition of cognitive skills form a combination that is hard to reconcile with rule-based models that often produce brittle behavior. Rule-based models can exhibit these properties by adhering to 2 principles: that the model gradually learns task-specific rules from instructions and experience, and that bottom-up processing is used whenever possible. In a model of learning perfect time-sharing in dual tasks (Schumacher et al., 2001), speedup learning and bottom-up activation of instructions can explain parallel behavior. In a model of a complex dynamic task (Carnegie Mellon University Aegis Simulation Program [CMU-ASP], Anderson et al., 2004), parallel behavior is explained by the transition from serially organized instructions to rules that are activated by both top-down (goal-driven) and bottom-up (perceptually driven) factors. Parallelism lets the model opportunistically reorder instructions, leading to the gradual emergence of new task strategies.

Keywords: Dual tasking, Psychology, Cognitive architecture, Complex systems, Human–computer interaction, Instruction, Learning, Situated cognition, Skill acquisition and learning, Knowledge representation, Symbolic computational modeling

1. Introduction

Rules and production systems are a popular choice of representation in modeling human cognition and have been widely successful in providing accurate and insightful accounts of human reasoning. However, such models have also been widely criticized. A first criticism is that models based on rules already have all the knowledge needed to solve the problem or do the

Requests for reprints should be sent to Niels Taatgen, Department of Psychology, Carnegie Mellon University, 5000 Forbes Avenue, Pittsburgh, PA 15213. E-mail: taatgen@cmu.edu

task (e.g., Dreyfus & Dreyfus, 1986; Rumelhart & McClelland, 1986). According to these critics, rule-based models may exhibit human-like performance on tasks, but fail with respect to their ability to offer an explanation of how such knowledge is acquired. For example, in a typical psychology experiment the participants are given instructions. Just on the basis of these instructions people are typically capable of doing the task. It requires quite a leap of faith to assume that instructions are automatically translated into rules. Many problems usually associated with insight problem solving also require the problem solver to go beyond what is stated in instructions, for example, by restructuring the problem. Another aspect of acquiring a task on the basis of instructions, which is hard to reconcile with the assumption that rules are learned directly from them, is that, although people can often perform the task based just on the instructions, they are typically slow initially and make many errors.

A second, related criticism of rule-based models is their "brittleness" (e.g., Holland, 1986): their behavior is limited to what their rules let them do. This may mean that if the model must perform a task that is similar, but not identical, to the original task, it may not work at all. This is an especially problematic issue when a model must operate in the real world, for example, in a robot or an information-foraging agent on the Web where unexpected situations can occur that are not covered by its rules. But brittle models also fall short of accounting for the flexibility and fine details of human behavior. The problem of brittleness has often been used as an argument against rule-based models as a viable approach to modeling human performance. For example, neural network models (e.g., Rumelhart & McClelland, 1986) can structure themselves on the basis of experience and need little knowledge at the outset. They can also adapt themselves surprisingly well to new situations, avoiding the brittleness problem. However, within this framework it is hard to model situations in which there is prior knowledge or a set of instructions to carry out.

We present an alternative view to problems with traditional rule-based models: It is not the rule representation itself that is problematic, but certain assumptions behind many rule-based models. Two assumptions in particular will be challenged in this article. The first is that, when participants in an experiment are given a new task, they almost immediately have access to a set of task-specific rules to perform it (the *instruction assumption*). Although this assumption is quite unreasonable, there are nevertheless many models that make it. As an alternative, following our earlier work (Taatgen & Lee, 2003), we employ models that gradually learn their rules during task performance on the basis of the instructions and background knowledge.

The second assumption of many rule-based models is that task representations are strictly hierarchical (i.e., Card, Moran, & Newell, 1983). The idea behind hierarchical task decomposition is that the task is decomposed into subtasks, each of which can be further decomposed into either deeper subtasks or sequences of primitive steps. Hierarchical task decomposition is based on sound software engineering principles and has been successful in describing task performance at higher levels, but it leads to brittleness at the finer levels of detail. The reason is that a strict hierarchical model enforces pure top-down control. Regardless of the situation in the world, the model follows its plan, even if it no longer makes sense given the context. Although brittleness can become evident in many ways, we focus on a category where it is particularly evident: multitasking situations, which can occur when either two or more tasks have to be done in parallel, or when different steps in a single task can partially or completely overlap in time (within-task multitasking). Such situations are particularly hard to model from a purely top-down perspective, because it assumes that it is possible to schedule all the components in the subtasks ahead of time. The alternative to top-down control is bottom-up control. Here the input from the environment is in full control. This approach is obviously not viable for all tasks because many require some form of internal state representation. To stress that a model should have as much bottom-up control as possible, we introduce the *minimal control principle,* which posits that humans strive for a strategy with a minimal number of control states. A control state marks where the strategy is in the task execution. In a fully hierarchical task representation every step in the process has its own control state, marking exactly where in the hierarchy the strategy is at that time. The disadvantage is that a fully flexible strategy should be able to address any possible event in any possible control state. The more control states there are, the larger the danger that there is some combination of a control state and an event that the strategy cannot handle, leading to brittle behavior. A strategy with a minimum number of control states produces a combination of top-down and bottom-up control, where the former is only used when needed, which is the case when two different situations cannot be distinguished on the basis of information that is available through bottom-up processes.

The general framework within which we develop these ideas involves the acquisition of skills in dynamic tasks in which timing and interaction with the outside world play an important role. This is a suitable domain for criticizing some conventional assumptions underlying rule systems, as well as for testing the adequacy of our own rule-based models. Skill acquisition has always been an important topic in cognitive modeling research, but the focus has been mainly on explaining speed improvements (e.g., Anderson, 1982; Logan, 1988; Newell & Rosenbloom, 1981; Taatgen & Lee, 2003). Another characteristic of a skill at the expert level that usually is not modeled is the reduced demand on attention. The classical example is that of a skilled driver having a conversation with his or her passenger, which is very hard for a novice driver. Various researchers have constructed cognitive models that can explain this ability to do things in parallel (Byrne & Anderson, 2001b; Freed, Matessa, Remington, & Vera, 2003; Gray, John, & Atwood, 1993; Meyer & Kieras, 1997), but these are expert models that do not learn and therefore rely, in some sense, on the instruction assumption. All of these models agree on the fact that parallel behavior can be achieved by the appropriate parallelization and interleaving of cognitive, perceptual, and motor actions, although there is still debate on whether cognitive actions can be carried out in parallel.

The goal of this article is to show how models of acquiring dynamic skills from instructions can make a start on the instruction and brittleness problems. The models are outfitted with task-general production rules that can operate instructions stored more or less literally. A relatively slow learning process then transfers the literary representation of the instructions into task-specific rules. We show that models with fewer control states, following the minimal control principle, produce more flexible skills and give a better account of the human data.

We first review some existing models of both skill acquisition and expert-level dual tasking. We then proceed to explain the model of a set of dual-tasking experiments by Schumacher et al. (2001), in which participants achieve perfect time-sharing between two simple tasks, along with the relevant aspects of the Adaptive Control of Thought–Rational (ACT–R) architecture (Anderson et al., 2004). We will see that the model of dual tasking learns a pure bottom-up strategy that translates into a model with just one control state. We then present a second model of a complex dynamic task, Carnegie Mellon University Aegis Simulation Program (CMU-ASP; Anderson et al, 2004), in which participants classify airplanes on a radar screen. This task has many opportunities for within-task multitasking and the usage of different strategies that can only be brought to bear if the task representation is sufficiently loose to allow exploration of these opportunities. Although this task needs at least some control states, the model presented here drastically reduces the number that would be required by a top-down model.

1.1. Cognitive models of skill acquisition

Skill acquisition involves progressing from slow, deliberate processing to fast, automated processing. Many models of skill acquisition assume that, by performing a task, learning processes transform the initial representation of the task into increasingly efficient representations. One of the first models proposed by Anderson (1982), based on Fitts's (1964) three-stage theory of skill acquisition, assumes that knowledge for a new skill is initially stored as declarative knowledge, using a simple propositional representation. Each declarative element is a symbol with labeled references to other elements. For example, the fact $3 + 4 = 7$ is represented by a symbol with references to "addition," "3," "4," and "7." Declarative memory can also be used to store a set of instructions. An example is a declarative instruction for a recipe, such as making tea: Put water in kettle, put water on stove until it boils, put tea leaves in teapot, pour boiling water in teapot, and wait 3 to 5 min. These five instructions for making tea can be stored almost literally in declarative memory. The simplicity of the representation explains why this is the starting point for a new skill: Declarative items of knowledge can be added as single items to memory. The disadvantage of declarative representations is that they cannot act by themselves; instead they need, according to Anderson's theory, production rules to be retrieved from memory and interpreted. This explains why initially processing is slow, because the declarative representations must be retrieved before they can be carried out, and it is prone to errors because the right declarative fact might not be retrieved at the right time. Another reason may be that initial knowledge is insufficient, so explicit reasoning is necessary to fill in the gaps. Finally, if declarative instructions are retrieved one at a time, it is often impossible to notice the potential for multitasking. In the tea recipe example, it would make sense to put the leaves in the teapot while you are waiting for the water to boil. A strict linear interpreter of the instructions would miss such an opportunity. While the model is processing declarative knowledge, several learning mechanisms start to gradually transform the declarative knowledge into production rules. In Anderson's (1982) original ACT* theory this was a set of mechanisms: proceduralization, composition, specialization, and generalization. One disadvantage of this theory was that these mechanisms together produced rules at such a rate that they could not be properly evaluated. This ACT–R is much more tractable because it has only one rule-learning mechanism, compilation, which we will discuss later. Once knowledge is transformed into a procedural representation, it is more efficient to use, because the inefficient retrieval step can be skipped.

Optimizing future efficiency on the basis of experience is the focus of two other learning models, chunking (Newell & Rosenbloom, 1981) and instance theory (Logan, 1988). Chunking, the learning mechanism employed by the SOAR cognitive architecture (Newell,

1990), learns new production rules, not from declarative knowledge, as in Anderson's theory (1982), but from a problem-solving trace that started with an impasse. The general idea is that novices begin with general problem-solving skills and domain knowledge, but not with appropriate rules to solve the problem right away. The novice uses these general skills to solve the problem, leading to search and slow performance. The chunking mechanism learns new production rules on the basis of these problem-solving attempts. These new rules can be used on future occasions where the same or a similar problem is encountered. A related model is provided by Logan's (1988) instance theory, which assumes two competing strategies for solving a problem: a general algorithm and a retrieval process. The algorithm is capable of solving the problem, in principle, but will take substantial time. An example is to solve an alphabet arithmetic equation such as $F + 4 = ?$ by counting four steps from F. Once the algorithm has solved a certain problem, the answer (e.g., $F + 4 = J$) is stored in memory. On future occasions the retrieval process tries to retrieve these past answers, called *instances* in Logan's theory. If $F + 4 = ?$ is presented again, the algorithm and the retrieval process simultaneously try to find the answer, and the strategy that finishes first wins the competition.

Our own previous work (Taatgen & Lee, 2002) concerns skill acquisition in ACT–R and also used the idea of gradually transforming declarative into procedural knowledge. In a model of the Kanfer–Ackermann Air Traffic Controller task (Ackerman, 1988), we demonstrated how an ACT–R model can learn the task from instructions. The model matched the experimental results on a global level (score per trial) and intermediate level (time per unit task), but it could not capture performance at the level of individual keystrokes. The problem was that landing each plane required some planning (determining which plane had to be directed to what runway) and then execution through a series of keystrokes. Participants in the experiments were perfectly capable of interleaving planning and execution, but the model could not, due to its hierarchical representation of the task.

Despite differences in representation and mechanisms, all four models share global similarities. In each case there is a transition from general task-independent methods to knowledge representations that are tailored to the task on the basis of experience. All four models produce a speedup in performance and a reduction in errors, as far as they are modeled, but they differ in whether and how they address other aspects of automaticity.

1.2. Models of parallel behavior

There are several theories on how multiple tasks can be carried out at the same time, but they share the same basic idea that the cognitive system consists of several modules that can each operate asynchronously and independently from each other. An early version of this idea is found in CPM-GOMS (Gray et al., 1993), which assumes that cognitive, perception, and motor actions are done in parallel, although within each modality actions are serial. This idea is also the foundation of the Executive-Process Interactive Control (EPIC; Meyer & Kieras, 1997) architecture, except that EPIC assumes that cognitive steps can be executed in parallel. Parallelization of behavior in these theories is a matter of optimizing the schedule of actions in all these modules. The APEX architecture (Freed et al., 2003) actually uses an algorithm to derive an optimal schedule. The peripheral modules of EPIC have been adopted into the ACT–R architecture (Byrne & Anderson, 2001b), but ACT–R continues to assume that central cognition is serial.

Parallel behavior is often studied in dual-task experimental paradigms. The dominant experimental paradigm is the *psychological refractory period* experiment (e.g., Pashler, 1994), in which participants do two tasks at the same time, but with the instruction to give priority to one task. The typical finding is that performance on the second task is slower than when the second task would have been carried out alone, indicating a cost of having to do both tasks at the same time. Schumacher et al. (2001) argued that this cost of doing two tasks at the same time may be an artifact of the priority given to one task. As a consequence, participants might postpone their response on the second task to make sure it is not made before the response on Task 1. To test this idea they asked participants to do two tasks in parallel with no order restrictions: a visual–manual task, in which a visual stimulus required a certain motor response, and an aural–vocal task, in which an aural stimulus required a certain vocal response. Both stimuli appeared at the same time, and the participant was instructed to react as fast as possible on both tasks. Schumacher et al. found that, given sufficient training, participants achieved perfect time-sharing, enabling them to do both tasks as fast as they would each task separately.

It is important to note that the usefulness of parallel processing is not restricted to multitask contexts. In many single tasks, such as the making-the-tea example, one can do several steps in parallel, and explaining this poses the same sort of questions as multitasking paradigms. Models that are based on parallel asynchronous modules have been quite successful in explaining various dual-task effects, as well as applied tasks such as menu and icon search (e.g., Byrne & Anderson, 2001a; Hornof & Kieras, 1997). The challenge is to explain how these efficient models can be learned.

1.3. Models of learning parallel behavior

Although instance theory does not deal with parallelism directly, it does account for why automated processes require less attention than controlled processes. According to Logan (1992), automation is characterized by diminishing attention requirements because, once attention is focused on a visual stimulus, the encoding and instance retrieval processes follow automatically and do not require attention, as opposed to the general algorithm. Despite the fact that instance theory sheds some light on the issue of attention, it is incomplete in the sense that it does not deal with the control issues that arise in complex skills. Kieras, Meyer, Ballas, and Lauber (2000) have such a theory, which posits five stages that the acquisition of multitasking must traverse, and they illustrated most of these with EPIC models. The stages are derived from analogies with how operating systems schedule multiple processes. In Stage 0, preprocedural interpretative multitasking (which they do not model), tasks are still encoded declaratively and must be retrieved and interpreted to be carried out. Stage 1 involves general hierarchical competitive multitasking, in which a general executive controls which task carries out its actions. While one task is performing its actions, the other tasks are "locked out" and are not allowed to carry out any step. In Stages 2 to 4, scheduling becomes increasingly more customized, eventually leading to a task-specific scheduler that is tailored to the specific task and that interleaves steps optimally. Although Kieras et al. specified the stages of learning, they did not provide a learning mechanism to accomplish it. Note that some of the problems to be solved in the EPIC model are unique to its hypothesis that central cognition is parallel, such as the need to "lock out" certain tasks. This implies that parallelism must be prevented early in the acquisition of a skill, as opposed to being learned later.

Chong and Laird (1997) modeled some aspects of learning dual tasking in their EPIC–SOAR model. Combining the perceptual–motor capacities of EPIC and the problem-solving capacities of SOAR (Newell, 1990), they provided a model that learns to solve conflicts in using resources. For example, if both tasks in a dual-task situation want to use the visual system, SOAR uses its problem-solving architecture to resolve this conflict, and learns new rules to avoid the conflict on future occasions.

Each of the three projects offers a partial solution to the learning problem. Logan's (1988) instance theory is consistent with the minimal control principle, because it reduces an algorithmic process with potentially many control states to an instance process with just one control state. However, it is restricted to situations in which a skill can be performed by a single retrieval from memory. Kieras et al.'s (2000) solution has the potential to decrease the number of control states, but lacks a learning mechanism. The EPIC–SOAR solution also cuts out some control states by learning rules that serve as shortcuts for longer chains of problem solving, but it is currently limited to solving resource conflicts.

2. A model of perfect time-sharing

This is our first example of a model that learns from instructions and uses the minimal control principle to achieve perfect time-sharing. We interleave the discussion of the model with an explanation of the relevant parts of the ACT–R (Anderson et al., 2004) architecture in which it is implemented.

Although ACT–R is in its roots a production system, it has evolved away from the classical paradigm. The latest developments are strong connections to functional MRI data (e.g., Anderson, 2005, this issue), reduction of the expressive power of production rules to make them more neurologically plausible, a more parsimonious mechanism for learning rules, and perceptual and motor modules from EPIC that interact with the outside world. Also, some traditional components of production systems have been removed, such as the use of a goal stack, because they produced predictions inconsistent with human data.

2.1. The central production system, modules, and buffers

ACT–R is designed to interact with software running the experiment or task. It can direct visual attention to a part of the screen and "see" what is there. Similarly, an ACT–R model can initiate a motor action such as pressing a certain key, which is then transmitted to the experimental program. This interaction with the world is carried out by specialized modules that make their results available to *buffers,* which are also used to issue new commands to the modules, such as the initiation of a key stroke or eye movement. Fig. 1 illustrates this concept. At the heart of this diagram is a production system. The rules collect their information not only from the perceptual–motor buffers, but also from other internal buffers and modules. Only the visual and the manual systems are depicted in the diagram, but ACT–R also implements an aural and vocal system. To retrieve a certain fact from declarative memory, a request must be made to declarative memory in the form of a partially completed pattern. The declarative module then tries to complete the pattern, after which the result is placed back in the retrieval

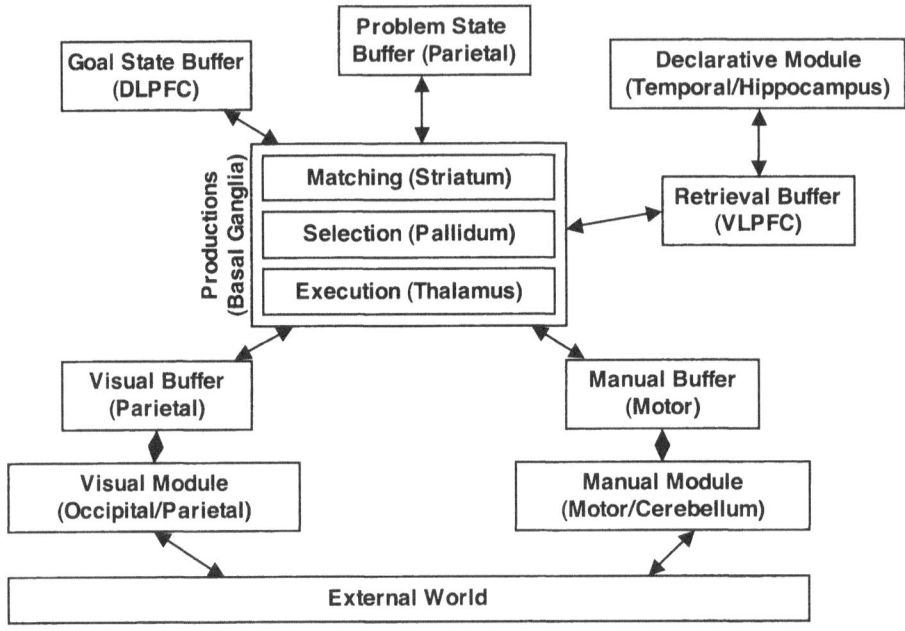

Fig. 1. Overview of the ACT–R architecture (adapted from Anderson et al., 2004).

buffer, where it can be matched and used by another rule. Another example is the visual system: To attend to a location, the production system must make a request to the visual system through the visual buffer. For our present discussion, the goal state buffer[1] is important, because it holds this control state. This means that the minimal control principle can be stated very precisely: *The model that is least susceptible to brittleness is the one with the fewest possible values in the goal state buffer.*

Production rules in ACT–R serve a switchboard function, connecting certain information patterns in the buffers to changes in buffer content, which in turn trigger operations in the corresponding modules. The advantage of this restricted power of rules is that modules can operate in parallel and relatively independent of each other. Behavior within a module is largely serial. For instance, the declarative model can only retrieve one item at a time, and the visual system can only focus its attention on one item in the visual field at a time. Timing within a module is usually variable and depends on the module. The central production system can also execute only one rule at a time, taking 50 msec per rule.

The perfect time-sharing experiments that are the topic of this section were reported by Schumacher et al. (2001). Participants performed a visual–manual task and an aural–vocal task concurrently. For the visual–manual task, a circle appeared in one of three horizontal locations to which they made a spatially compatible response with their right hand, pressing their index, middle, or ring finger to left, middle, or right locations. For the aural–vocal task, participants were presented with tones that lasted 40 msec and were either 220 Hz, 880 Hz, or 3520 Hz, to which they responded "one," "two," or "three." The experiment consisted of homogeneous single-task blocks, in which participants did just the visual–manual task or just the aural–vocal task, and mixed-trial blocks. The mixed-trial blocks consisted of dual-task trials, in which both stimuli were presented simultaneously, and heterogeneous single-task trials, in which only one of the two stimuli were presented. The participant would therefore never know whether just one or both stimuli would be presented in a trial.

Byrne and Anderson (2001b) modeled this experiment at the level of expert behavior. Instead of presenting their model, we walk through the expert level of our own model, which is almost identical. Fig. 2 gives a time diagram of the model's execution. The left-hand side represents the moment that both stimuli are presented, and time progresses to the right. A block in the diagram represents processing in one of the modules and arrows indicate dependencies between them.

1. First, the visual module notices that a new stimulus has appeared on the display. It responds to this by updating the visual buffer, noting that a new stimulus is in the visual field, and giving its location. This ACT–R assumption is that this is instantaneous, so there is no box in the diagram.
2. A production rule (compiled-attention-rule) notices the update in the visual buffer and puts a request in the visual buffer to attend the new visual stimulus. At the same time the aural module detects the tone and updates the aural buffer to indicate this.
3. A production rule (attend-aural) puts in a request to identify the tone that has been detected by the aural module.
4. The visual module finishes the eye movement and encoding of the visual stimulus, having found the pattern "O – –," which it places in the visual buffer.

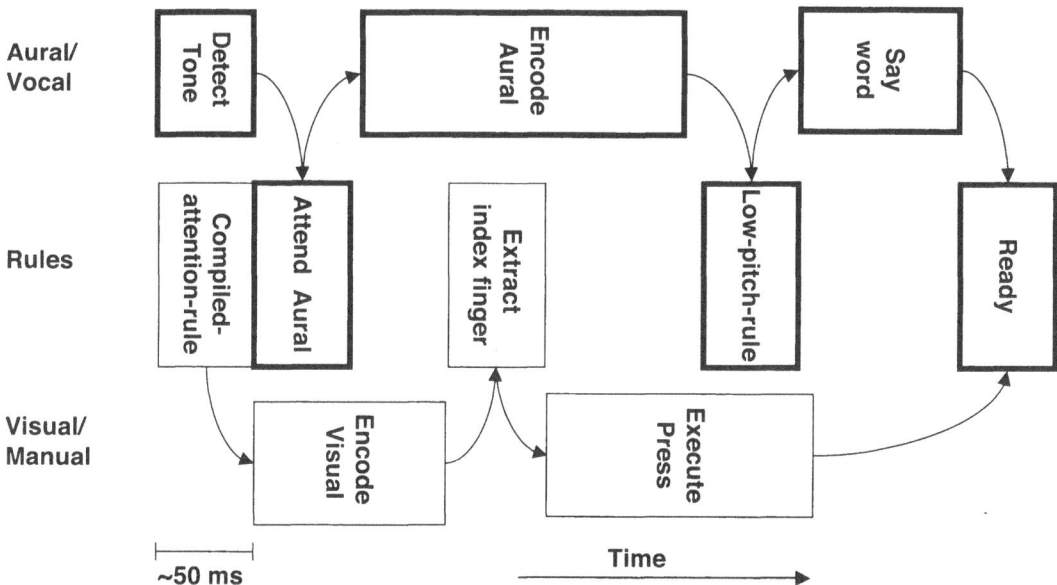

Fig. 2. Time diagram of expert-level parallel behavior in the Schumacher et al. (2001) Experiment 1 according to the ACT–R model. Arrows indicate dependencies between steps. Thin-bordered boxes represent steps in the visual–motor task and thick-bordered boxes in the aural–vocal task.

5. A production rule (extract-index-finger) notices the "O – –" pattern in the visual buffer and puts in a request to press the index finger in the manual buffer.
6. The motor module notices the request in the manual buffer and starts initiating the key press.
7. The aural module finally determines that the tone is a low pitch and places this information in the aural buffer.
8. A production rule (low-pitch-rule) notices that a low pitch has been detected and puts a request to say "One" into the vocal buffer.
9. The vocal module starts processing the request to say "One."
10. When both the vocal and the module are done, the rule "Ready" terminates the trial.

In the timing diagram, the thin-bordered blocks are steps in the visual–manual task, whereas the tick-bordered blocks are steps in the aural–vocal task. There are no gaps between blocks of each of the two tasks, indicating that the time-sharing is perfect.

2.2. The role of declarative memory

Up to this point we have discussed the contributions of the perceptual–motor systems and the central role of the rules. To explain how the highly specialized rules can be learned, two additional components are needed: declarative memory and a mechanism to learn new rules from its contents. Declarative memory holds information in a fairly literal form. It can be used to store all kinds of knowledge, including facts (such as $3 + 4 = 7$), old experiences (the light switch in the closet is not operational), and instructions (boil water to make tea). Declarative knowledge is passive: It is stored and will only be brought to bear if it is requested by a production rule action. The process is similar to the interaction with the perceptual–motor systems: A production rule issues a request to declarative memory for a certain fact by placing part of the fact as a cue in the retrieval (declarative) buffer (i.e., $3 + 4 = ?$, the light switch in the closet ?). The declarative memory module then tries to complete the pattern. This completion attempt fails if no matching memory is available or if the matching memory pattern has decayed too much. The exact characteristics of declarative memory in ACT–R can be found in Anderson et al. (2004). What is important for this discussion is that strong memories are recovered very quickly from declarative memory, whereas weak memories are recovered more slowly or not at all.

The assumption for the model of the Schumacher task (Schumacher et al., 2001) is that the instructions are stored in the following declarative form:

1. Attend any aural stimulus.
2. Attend any visual stimulus.
3. Given an attended aural stimulus, retrieve and make a response for that stimulus.
4. Given an attended visual stimulus, determine and make a response for that stimulus.

The assumption in these instructions is that to process a stimulus, either visual or aural, it first has to be attended, that is, the pitch of the tone or the identity of the visual stimulus must be determined. This is taken care of by Instructions 1 and 2, whereas Instructions 3 and 4 prescribe what should be done with a stimulus once it has been identified. For the aural stimulus, this requires retrieving a mapping from declarative memory (i.e., low-pitch maps onto "one") and then giving that response. For the visual stimulus, there is a straightforward mapping between location and finger. We therefore assume that no memory retrieval is necessary to determine this mapping.

As declarative knowledge is passive by itself, the instructions given previously must be retrieved, interpreted, and carried out by production rules. This is much slower than the process with the expert model. This has consequences for the optimal order in which the instructions should be carried out, because as long as the model is in the instruction interpretation stage, dual-task costs are incurred regardless of the execution order. It is therefore almost impossible to initially determine the optimal order, because this might change later due to learning. Also, given that some of the trials in a mixed block are single-task trials, not all of the instructions necessarily have to be carried out on each trial, frustrating a plan that waits for a stimulus that never appears. Here we invoke the minimal control principle: Instead of having states for the different stages in the process, one state is sufficient, because the instructions themselves contain information on when they are applicable. The only thing that is needed are interpretation rules that retrieve the right instruction at the right time.

To retrieve the proper instruction at the right moment, there are four retrieval rules for each of the possible event types, such as

Get-next-visual-location-instruction-rule
IF   the goal is to do a certain task (goal buffer)
AND   the visual module has detected a new stimulus (visual buffer)
THEN   request an instruction to do something with a visual stimulus in the context of the current task (retrieval buffer)

A second rule then executes an instruction, such as the one to attend the visual stimulus:

Attend-visual-stimulus
IF   the goal is to do a certain task (goal buffer)
AND   the instruction is to attend a visual stimulus (retrieval buffer)
AND   there is a location in the visual buffer (visual buffer)
THEN   put a request in the visual buffer to attend that location (visual buffer)

In total, the model has seven execution rules, one to attend to a visual stimulus, one to attend to an aural stimulus, one rule to retrieve a pitch-word combination, one rule to say the word that has been retrieved, and three rules that implement the mappings of the visual stimuli on the three fingers that are used to make the response. The assumption is that because the stimulus-response mapping between the visual stimuli and the response is compatible, no retrieval from declarative memory is necessary. The following encoding parameters were estimated: 50 msec to recognize that there is a tone (which is the ACT–R default), 200 msec to determine the pitch of the tone (default is 285 msec), and 100 msec to attend the visual stimulus (default is 85 msec). For verbal and motor responses, ACT–R's default values were used. Fig. 3 shows a diagram of the behavior of the model as a novice. Each step consists of at least three substeps: a rule that retrieves the next instruction, the retrieval of the instruction, and the execution of the retrieved instruction. The steps that are part of the visual–manual task are in the thin-bordered boxes. We can derive that some dual-task costs occur from

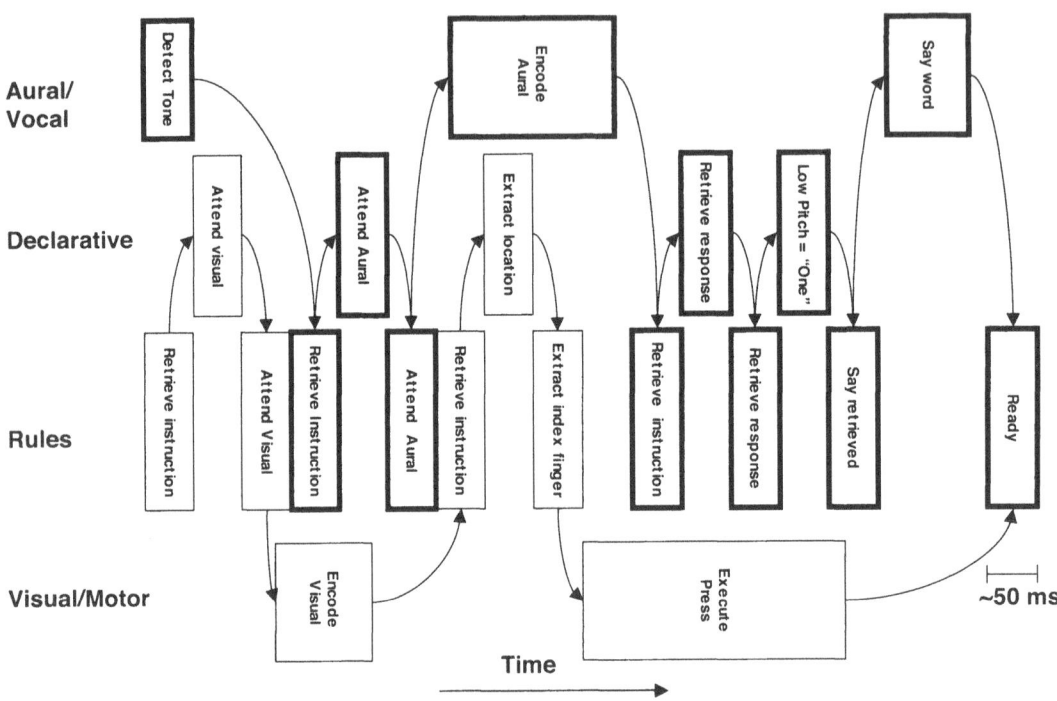

Fig. 3. Time diagram of a novice in the Schumacher et al. (2001) Experiment 1 according to the ACT–R model.

the fact that not all thin-bordered boxes are horizontally next to each other: the time between the encoding of the visual stimulus and the retrieval of the instruction that extracts the position of the circle from that stimulus. The aural–vocal task (represented by thick-bordered boxes) also has a dual-task cost because it must wait for completion of the initial steps that handle the visual stimulus.

2.3. From declarative to procedural knowledge: Production compilation

To explain how novice behavior can develop into expert behavior through training, one final component of ACT–R must be described: *production compilation,* the mechanism for learning new production rules. Production compilation (Taatgen & Anderson, 2002) learns new rules by combining two existing rules that fire in sequence into one new rule. If the first of the two rules makes a request to declarative memory, the result of which is used by the second rule, then the retrieved chunk is substituted into the new rule, effectively eliminating the retrieval. In itself this mechanism only produces more efficient and specialized representations of knowledge that are already available, which is sufficient for our present purpose. However, when the production rules that are compiled are a general cognitive strategy, the resulting rules, although specializations of the general strategy, nevertheless generalize the specific experience.

An example of production rule learning is the combination of the rule *Get-next-visual-location-instruction-rule* and *Attend-visual-stimulus* shown earlier. When they are compiled with the instruction to attend any visual stimulus, ACT–R learns the rule:

Compiled-attention-rule
IF   the goal is to do the Schumacher task (goal buffer)
AND   the visual module has detected a new stimulus (visual buffer)
THEN   attend that stimulus (visual buffer)

Newly learned rules can be the source of recombination themselves, enabling the collapse of more than two rules into one. For example, the following rule summarizes the processing from three rules and two retrievals, and it was learned by first compiling the first two rules with the retrieval of the instruction and then compiling the resulting rule with the final rule and the retrieval of the response set (low-pitch—"one"):

Learned-Low-Pitch-rule
IF   the goal is to do the Schumacher task (goal buffer)
AND   a tone with a low pitch has been detected (aural buffer)
THEN   say "one" (vocal buffer)

This rule, like the earlier compiled rule, is state independent: It reacts to the fact that a tone with a low pitch has been detected. As a result, the rules that make up expert behavior produce reactive, bottom-up behavior: Steps are not planned but cued by the environment.

After all the steps in the novice model have been combined, the behavior depicted in Fig. 2 is reached. The expert behavior shown in Fig. 2 involves no more dual-task costs, because there is no interference between the two tasks. Of course, not all combinations of rules can be compiled. If a first rule makes a request to a perceptual module and the second rule acts on the result from the module, the rules cannot be compiled because the outcome depends on external input. Similarly, if two rules send a request to the same module (e.g., two consecutive key presses), they also cannot be compiled. As a result, learning eventually produces rules that move the system from one perceptual input or motor output to the next (as far as they depend on each other), making perceptual and motor actions the main determiner of performance. This model indeed reaches this stage (Fig. 2), but if a task is more complex, there may never be enough time to learn all the possible rules that are needed.

2.4. Gradual learning of production rules

If we were to introduce these new rules right away into the system, expert behavior would be reached in a few trials, which obviously does not occur with humans in the Schumacher task (Schumacher et al., 2001). Anderson, Fincham, and Douglass (1997) also found evidence that procedural learning is a slow process. To account for this finding, instead of introducing these new rules into the production system right away, ACT–R gradually phases them in by using *production utility,* the subsymbolic component of production rules. Utility is calculated from estimates of the cost and probability of reaching the goal if that production rule is chosen, with the unit of cost being time. ACT–R's learning mechanisms constantly update the parameters used to estimate utility based on experience. If multiple production rules are applicable for a certain goal, the production rule with the highest utility is selected. This selection process is noisy, so the rule with the highest utility has the greatest probability of being selected, but other rules have an opportunity as well. This strategy may produce errors or suboptimal behavior, but it also lets the system explore knowledge and strategies that are still evolving.

When a new rule is learned, it is given a very low initial estimate of utility. This value is so low that the new rule effectively has no chance of being chosen, because it must compete with the parent rules from which it was learned. To have a reasonable chance of selection, the rule must be recreated a number of times. This has the advantage that only new rules which correspond to frequent reasoning patterns will make it into the effective rule set. More formally, utility has two components, one that it inherits from the parents and another based on the experiences that the rule itself gains, combined in the formula

$$U_p = \frac{mU_{p,\text{prior}} + \text{experiences} \cdot U_{\text{experienced}}}{m + \text{experiences}}$$

In this equation, $U_{p,prior}$ represents the utility component inherited from the parent rule, which starts out very low but which is increased at each recreation of the rule, and $U_{experienced}$, the average of the experienced utility of the rule (which will only have a nonzero value if there have been experiences). Each component is scaled, $U_{p,prior}$ by m, a fixed parameter that is set to 10 by default, and $U_{experienced}$ by the number of actual experiences. As a consequence, the impact of $U_{p,prior}$ is large initially, but decreases in importance as the rule gains its own experiences. $U_{p,prior}$ is set to -20 when the rule is first created, but after each recreation its value is increased toward the value of its parent's utility according to the equation

$$U_{p,prior}(t+1) = U_{p,prior}(t) + \alpha[U_{parent}(t) - U_{p,prior}(t)]$$

The new value of $U_{p,prior}$ is made equal to its old value plus a fraction of the difference between its current value and the utility of its parent. As a result, the utility value of a new rule will initially increase toward its parent's value quite steeply, but then levels off as it approaches the parent's value. However, as it approaches the parent's value, the new rule gets a reasonable chance of winning the competition and gaining experiences itself, because selection is a noisy process.

Fig. 4 gives an example of how this works in practice based on the parameters used in the model, depicting the growth of utility for a new rule and the competition with its parent. In this example, the utility of the parent rule is 10, and the actual utility of the new rule is 10.1, indicating a 100 msec gain (it saves one production rule and one retrieval), but the model can only discover this by having experience with the rule. The x axis represents opportunities for the new rule to fire, with the creation of the rule at 0. The rule starts with a utility of -20, and thus has zero probability of winning the competition with its parent. When the parent wins, however, the new rule is recreated, increasing its utility. At around 100 recreations, the new rule starts having a small probability of winning the competition and therefore gaining its own experiences. At some later point, the new rule's utility surpasses that of its parent due to experiences with a utility of 10.1, eventually making it the dominant rule. In terms of reaction times, this particular rule produces a speedup of 100 msec that starts building up at around 100 experi-

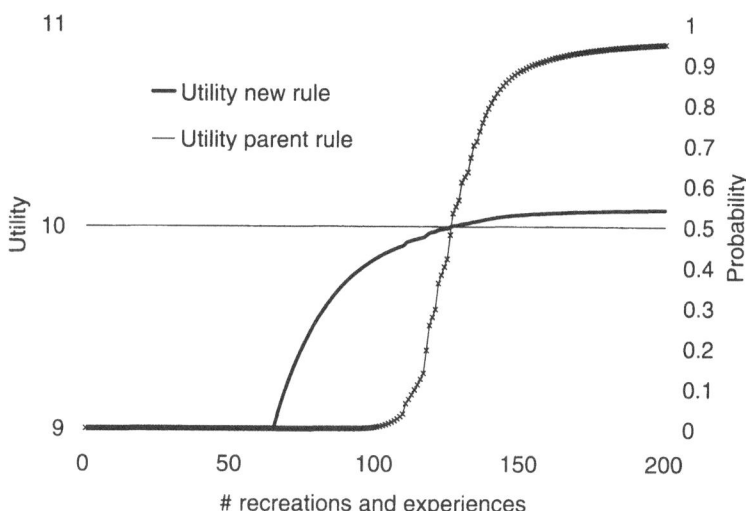

Fig. 4. Example of gradual introduction of a new production rule. The thin line represents the utility of the parent rule, and is fixed on 10. The new rule starts with a utility of −20 represented by the thick line (which is initially outside the graph's scale), but eventually approaches its true value of 10.1. The S-shaped curve represents the probability (with its axes on the right side of the diagram) that the new rule will win from the parent rule.

ences and reaches a peak around 150 experiences, where it wins from its parent around 95% of the time. This mechanism of gradual introduction of new rules produces a slow transition from the behavior in Fig. 3 to the behavior in Fig. 2. A consequence of the gradual introduction mechanism is that there is always a small residual probability that the parent is selected instead of the new rule, occasionally producing a suboptimal trial. We now discuss the three Schumacher et al. (2001) experiments in detail and compare the model's outcome to the data from the participants.

2.5. Experiment 1

In Schumacher et al.'s (2001) first experiment, pure-trial blocks consisted of 45 trials each, whereas mixed-trial blocks consisted of 48 trials, 18 dual task and 30 single task. On Day 1 of the experiment, participants were trained just on pure-trial blocks, 6 of each type. Day 2 consisted of 6 pure-trial blocks and 8 mixed-trial blocks, and Days 3 to 5 consisted of 6 pure-trial blocks and 10 mixed-trial blocks. Fig. 5 shows the reaction times for each of the subtasks of the experiment for both participants and the model. On Day 2 there is a clear dual-task cost for the aural–vocal task, because it is slower than both single-task variants. The dual-task cost for the visual–manual task is almost nonexistent, suggesting that, on Day 2, participants complete the visual–manual task before the aural–vocal task. On Day 5, however, all dual-task costs have disappeared. The model nicely replicates all these effects and also reaches a stage in which no dual-task costs are incurred. The residual probability that a parent rule is selected instead of the optimal rule is small in this model (<1%), and even then it does not affect dual-task costs because there is still some "slack" in Fig. 2 to compensate for it.

Visual-manual task

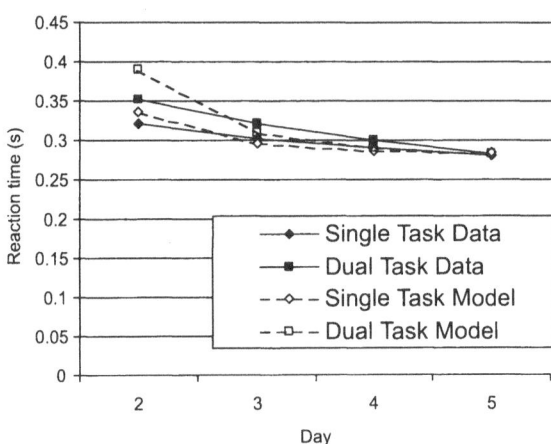

Fig. 5a. Results of Experiment 1 of Schumacher et al. (2001), model and data, visual-manual task. Single-task trials are averages of homogeneous and heterogeneous trials.

Aural-vocal task

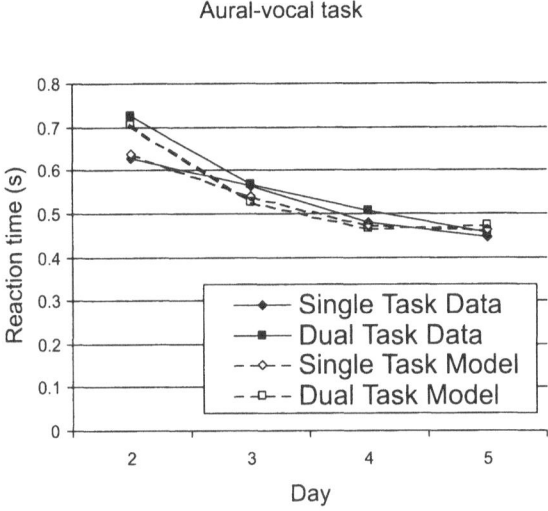

Fig. 5b. Results of Experiment 1 of Schumacher et al. (2001), model and data, aural-vocal task. Single-task trials are averages of homogeneous and heterogeneous trials.

2.6. Experiment 2

According to Schumacher et al. (2001), dual-task costs in psychological refractory period experiments are often due to ordering instructions. To further show this, they took the participants from Experiment 1 and gave them additional blocks of trials, but now with the explicit instruction to always give the vocal response before the motor response. The visual stimulus was presented

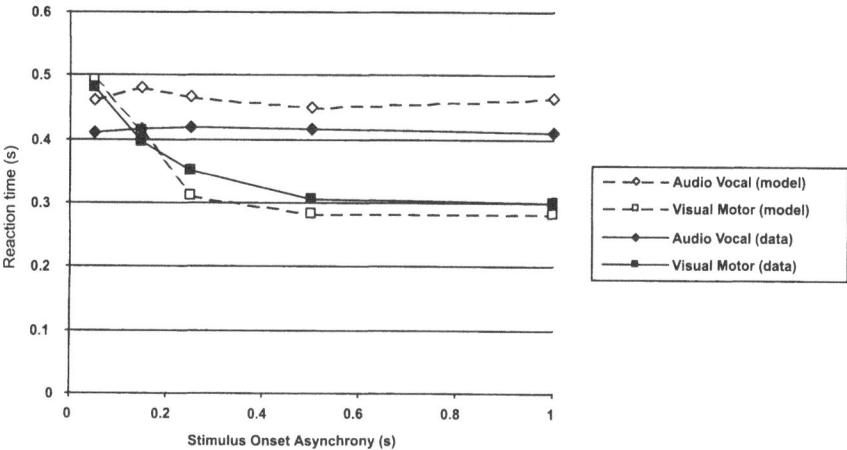

Fig. 6. Results from Experiment 2 of Schumacher et al. (2001), model and data.

50, 150, 250, 500, or 1,000 msec after the aural stimulus, the stimulus onset asynchrony (SOA) that is typical for psychological refractory period experiments. The model achieved this by introducing a single top-down constraint to the rules that issue the motor commands, by specifying that the manual response could only be initiated after the initiation of the vocal response. Fig. 6 shows the results of the experiment and of the model simulation. The primary task, the aural–vocal task, shows no effect of SOA and has no additional costs due to the second visual–motor task. The visual–motor task shows dual-task costs for SOAs of 250 msec and smaller, because the manual response must wait until the vocal response has been initiated.

2.7. Experiment 3

Experiments 1 and 2 have made plausible that dual-task costs can at least sometimes be attributed to instructions. However, Schumacher et al. (2001) were looking for evidence consistent with the absence of a central bottleneck, that is, for the hypothesis that processing in central cognition occurs in parallel. Because the visual–manual task is so much easier than the aural–vocal task, serial central processing could not be ruled out. Indeed, the ACT–R model has a central cognitive bottleneck and is perfectly capable of modeling the data. To make the visual–motor task harder, they added one extra stimulus and made the mapping incompatible. Now there were four possible positions for the visual stimulus, and the mapping of the positions on the finger was reversed—for example, for the right-most position the left-most finger had to be pressed. The experiment was otherwise identical to Experiment 1, except that it was extended to 6 days. We modified the model to accommodate the new task. First, determining which finger corresponds to what location can no longer be considered automatic, so we changed the instruction to determine the finger to be pressed to retrieve a mapping between location and finger from memory. Second, because the response selection moments are much closer in time than in Experiment 1, we added noise to the perceptual encoding times to make more accurate predictions, so the time to determine the pitch of the tone varied from 0.1 to 0.3 sec, and the visual encoding time varied from 0.1 to 0.2 sec. Finally, we increased the duration of the production rule that initiates the motor output from 0.05 to 0.2 sec. The reason for this last change was that incompatible mappings have a lasting effect on the response time that the model cannot explain. Although we have some ideas on how to accommodate this matter, it is not the topic of this article, so we modeled the difference parametrically. The results (Fig. 7) show that, in both the model and the experiment, there are some residual dual-task costs, although the model only shows them in the aural–vocal task. A further analysis of Schumacher et al. showed that 5 of the 11 participants had virtually no dual-task costs by the end of the experi-

Aural-Vocal task

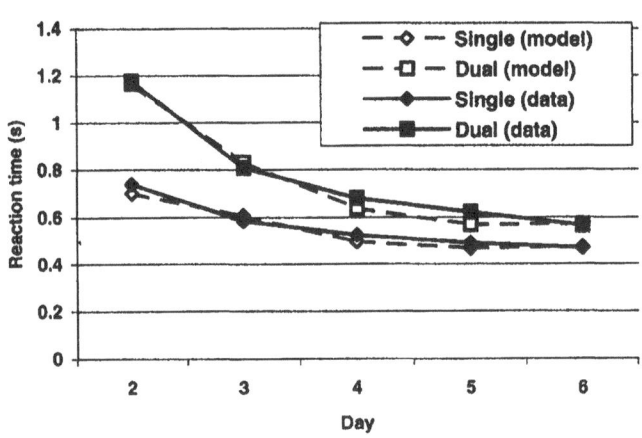

Fig. 7a. Results of Experiment 3, model and data, aural-vocal task. Single-task trials are averages of homogeneous and heterogeneous trials.

Visual-motor task

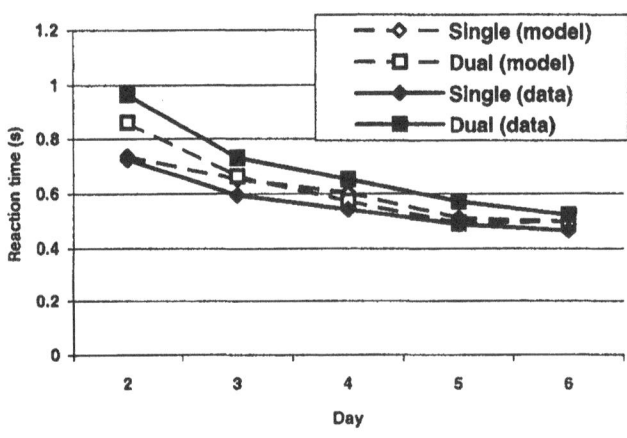

Fig. 7b. Results of Experiment 3, model and data, visual-motor task. Single-task trials are averages of homogeneous and heterogeneous trials.

ment. These participants were also the fastest on the task. When the duration of the production rule that initiates the motor response is lowered from 0.2 sec to 0.15 sec, the model reproduces this effect. Also, when this value is increased to 0.35 sec, the model produces the behavior of the 4 participants who had large dual-task costs (Fig. 8). This suggests that the ability to handle the incompatibility in the mapping between the visual stimulus and the finger to be pressed can explain the observed individual differences. However, more solid evidence is needed to corroborate this finding.

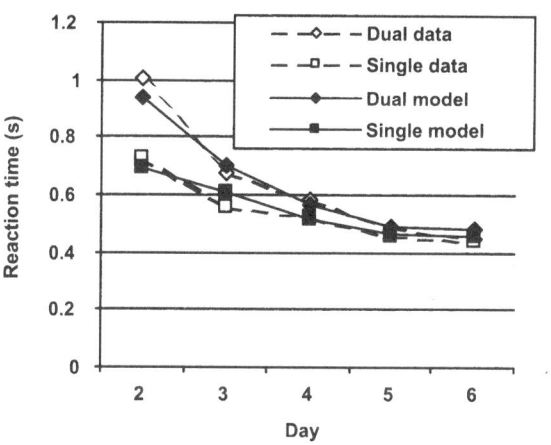

Fig. 8a. Data and model for 5 participants with no dual-task costs (low interference). Results are averaged over the two tasks and the two types of single-task trials.

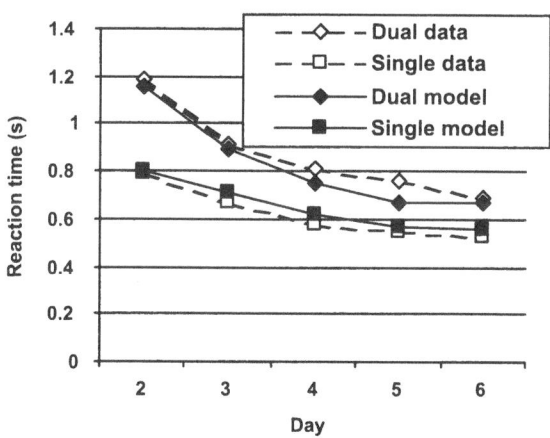

Fig. 8b. Data and model for 4 participants with high dual-task costs (high interference). Results are averaged over the two tasks and the two types of single-task trials.

2.8. Discussion

The models of the Schumacher et al. (2001) experiments start out with a set of instructions that must be retrieved when needed. Eventually they learn production rules that are sensitive to the conditions under which they should be applied, which are usually independent of order, with the exception of Experiment 2 in which the motor response must be made before the vocal response. This order independency is an important aspect of the ability to parallelize behavior. ACT–R's central cognitive processes are serial, but firing a single production rule only takes 50 msec. If all relevant knowledge is fully proceduralized and ready to respond to the conditions in which it is applicable, this restriction will hardly ever be the bottleneck of processing. When knowledge is still declarative, the duration of the bottleneck increases, because retrieving and carrying out an instruction takes a significant amount of time.

It would have been very hard to model this task with a more hierarchical representation of instructions with top-down control, unless an explicit general executive is added that schedules all the steps. The proper order of carrying out the instructions would then be a matter of planning, which can only be done when the instructions are proceduralized, because only then can it be determined how much time they take. This model is driven purely by the perceptual inputs and ends up with a greedy strategy that chooses any step that it can carry out at the moment. Although a greedy planning strategy is not optimal in general, it is for this task. We can therefore conclude that the minimal control principle is useful for a small task, both in describing human behavior and in prescribing how a model of that behavior should be constructed. Whether a greedy strategy is always the best solution, both in the descriptive and prescriptive sense, remains to be seen.

Hazeltine, Teague, and Ivry (2002) conducted a further series of dual-task experiments in an effort to rule out a central bottleneck. With models similar to the ones discussed here, Anderson, Taatgen, and Byrne (in press) demonstrated that a central bottleneck model can still model these more intricate experiments and can even explain subtle dual-task effects in some of them.

3. Skill acquisition in complex dynamic tasks

Principles derived from simple tasks should scale up to real-world cognition, and we therefore shift our attention to a real-time, complex dynamic task. Although it does not involve any explicit multitasking, there are many opportunities for what could be called *within-task multitasking*. In most of the examples we examine, this within-task multitasking consists of two instructions that are initially carried out in sequence, but that start overlapping with experience.

The CMU-ASP task is a simplification of Georgia Tech Aegis Simulation Program (Hodge et al., 1995), in which the participant takes the role of an operator on a ship who must use a radar screen that displays various tracks of airplanes. The goal is to classify these planes, which involves asking the system for information and entering the classification by a sequence of keystrokes. Fig. 9 shows the interface to the system. The circular area that makes up most of the screen is the radar scope. The small circle in the middle represents the home ship, and the rect-

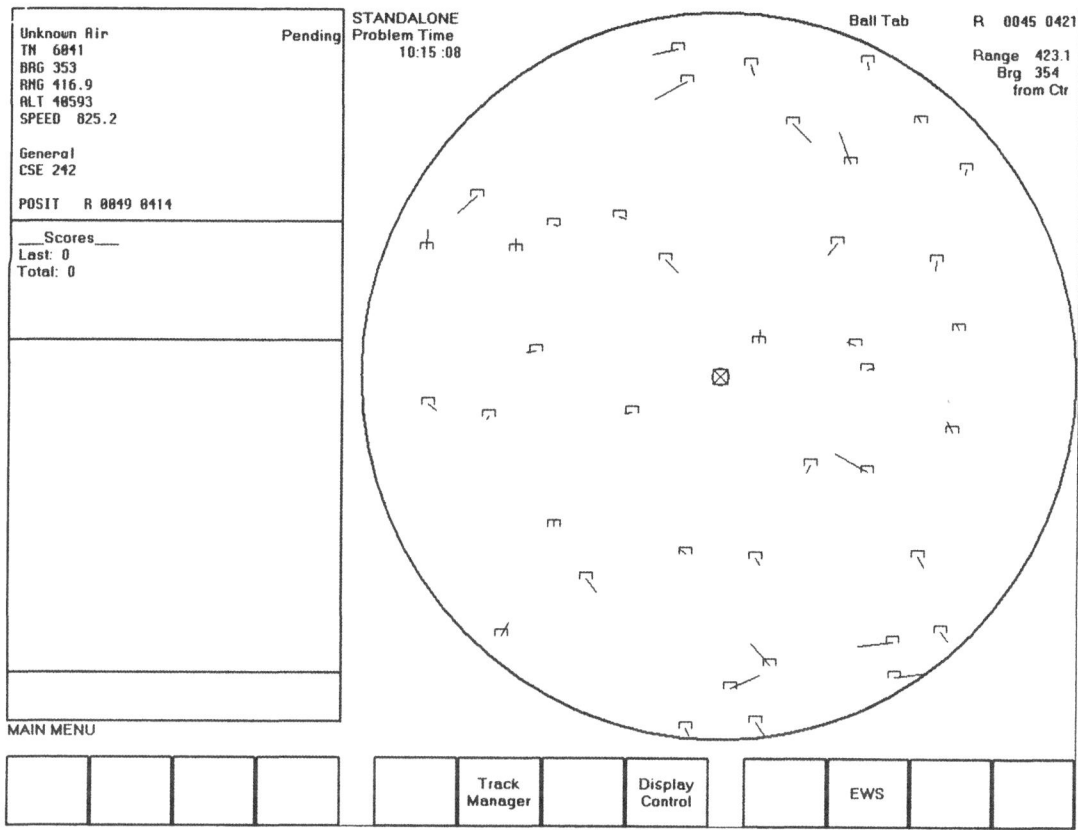

Fig. 9. The CMU-ASP task.

angles with lines emanating from them represent unidentified planes. Participants can select a plane to work on by clicking it with the mouse, after which some information about that plane will appear in the top-left area of the screen. The information important to this task is the altitude (ALT) and speed (SPEED), which are sometimes enough to determine the identity of a plane. A third region of importance is the bottom row of 12 boxes, which correspond to the 12 function keys on the keyboard. For example, to activate the Track Manager, a menu choice to enter airplane classifications into the system, the F6 key must be pressed because the label "Track Manager" is in the sixth box from the left. Labels change dynamically depending on the context: Once F6 has been pressed, a new set of functions is displayed in the bottom row. Because the labels in the bottom row change, mastery of this aspect of the task requires considerable learning.

To do the task, participants must take three steps for each plane. The first involves selecting a plane to work on, which is accomplished by clicking on one of the unidentified planes on the radar screen. Although any unidentified plane will do, the instructions specify some preference relations: Planes that are close to the home ship (the middle of the scope), planes that are fast (as identified by length of the line emanating from the rectangles), and planes that are headed for the home ship (when the line from the rectangle is pointing toward the center circle) should be classified first. Step 2 concerns determining the identity of the plane, for which there are two methods both of which are only sometimes successful. The first method is to look at the speed and altitude in the top left of the screen. If speed and altitude are within certain margins, the plane is a commercial airplane, otherwise the identity cannot be determined by that method. The second method is to use radar profiling, which is accomplished by giving the "EWS" command (by pushing F10) and confirming this command with a second key press. Radar profiling may give an answer that allows a direct classification, or it may return "negative." If neither classification method is successful, the participant should move to the next plane, otherwise they must proceed with Step 3: Enter the identity into the system via the Track Manager, which requires a sequence of eight key presses. For example, to enter a commercial airplane, one enters function keys corresponding to the labels: "Track Manager"—"Update hooked track"—"Class/Amp"—"Primary ID"—"Friend"—"Air Id Amp"—"Nonmilitary"—"Save Changes". For each successful and correct classification, points are awarded that are displayed in the middle left of the screen. The number of points depends on the priority of the plane and the time within the scenario that it is classified: A fast plane that is close by and headed toward the home ship will yield many points if it is classified early, but when it is classified later its score will be smaller. The scoring scheme encourages participants to identify the important planes early in the scenario.

Anderson et al. (2004) reported an experiment that consisted of 20 scenarios of 6 min each, divided over 2 days. Participants showed considerable improvement in both speed and the number of classified planes. The results can be found in Fig. 11, which we discuss along with the model's ability to match participants' behaviors.

3.1. From a strong to a weak hierarchy

Anderson et al. (2004) modeled learning on the CMU-ASP task using a declarative representation of instructions in combination with production compilation. The model uses a hierarchical decomposition of the task into instruction sequences that can be interpreted by production rules. Each sequence consists of a number of steps that are carried out in order, some directly and others by referring to another rule for which a subgoal is created. The model learns by compiling the retrieval of instructions into task-specific rules and by learning which labels correspond to which keys.

Although the Anderson et al. (2004) model is successful in modeling the performance improvement in terms of speedup at the level of the three subtasks, and it can also predict how long attention is focused on certain regions of the screen during these subtasks, it is limited by the representation of instructions. It can model the speedup in performance but not the additional flexibility and parallelism that are associated with skilled behavior. In contrast, we present an updated model that accounts for three examples of additional flexibility:

1. Selection of a plane. Although the instructions state which planes should be classified first, participants initially do not seem to make much of an effort to optimize their selection. As they gain more experience, they gradually pay more attention to this, resulting in an improvement in the quality of the selected planes. The model explains this shift by dual-tasking the instructions to find a plane and clicking on a plane.

2. Selection of the appropriate function key. With practice, participants no longer have to look at the function key labels at the bottom of the screen, but rather know which to press. The Anderson et al. (2004) model explains this by recalling past goals, but we provide an alternative explanation in which the dual-tasking of finding a label and finding the corresponding function key makes label finding obsolete.

3. Optimization of hand movements. Interaction with the task requires using both the keyboard and the mouse, which means that the right hand must be moved between the two. However, moving the hand at the moment that it is needed is inefficient. Instead, its use can be anticipated, so that, for example, the hand is on the mouse at the moment that a new plane must be selected. Another option is to use the left hand for the keyboard and the right hand for the mouse.

3.2. Minimizing control

In Anderson et al.'s (2004) CMU-ASP model, the cue for retrieving the next instruction is the instruction that has just been carried out. This creates a strict top-down control of behavior, because the model is always driven by internal representations. Every step in the task execution therefore corresponds to a control state, effectively maximizing the number. Our previous dual-task model breaks the sequencing of instructions by having only a single control state and by using perceptual events as cues, producing pure bottom-up control. However, the complexity of the CMU-ASP task does not allow for pure bottom-up control. To be able to model flexibility in the CMU-ASP task, a compromise between bottom-up control and top-down control is required, because the model not only must react to perceptual and motor events, but also keep track of where it is in the process. Our solution is to replace the strong hierarchical instruction structure with a weak one with as few control states as possible, again following the minimal control principle. More specifically, it consists of three components:

1. Instead of a hierarchical structure, instructions are organized in sets that correspond roughly to unit tasks. Each instruction set has one or more conditions that specify when it is applicable. Within each set, instructions are sequenced by pairwise association strengths. Association strength is a feature of ACT–R's declarative memory not yet explained that increases the activation of a declarative memory element associated with this goal. We assume that associations between subsequent instructions are learned while reading and memorizing the instructions. Once the model starts carrying out the instructions, the next instruction in the sequence will receive extra activation because its predecessor has just been executed.

2. The instruction retrieval mechanism tries to retrieve an instruction within this set, relying on ACT–R's activation mechanisms to access the right instruction, usually the next one. Other sources of activation can influence the selection as well, such as a perceptual event that activates an instruction to handle that event type.

3. The rules that carry out a retrieved instruction check whether any constraints on the instruction are met, because the retrieval process does not guarantee this. For each instruction, it may take several steps to carry them out depending on the situation. For example, an instruction to click on a plane will initiate a movement of the hand to the mouse if the hand is not already there. If the hand is on the mouse and the mouse pointer is on the plane, then it will immediately initiate a click.

Within this weak hierarchical representation, each instruction can be considered as a control state, because it more or less determines the next step. However, once instruction retrieval has been compiled into task-specific rules, the instructions are no longer retrieved; the control states dissolve and leave only a single control state for the entire rule set. As a consequence, once the model has compiled all the instructions into rules, it has as many control states as rule sets, giving a large improvement over the strong hierarchical representation, which has as many control states as individual instructions.

3.3. The CMU-ASP model and global results

The model of the task uses declarative instructions that are organized into the five rule sets outlined in Fig. 10. The first four sets correspond to the main unit tasks: Select a plane, use one of two strategies to find the classification, and then enter the classification into the system. Each of the sets has one or more preconditions (in the hexagons) that must be satisfied before the set itself is activated. Only one set is active at a time, which is represented in the goal state buffer, but in principle it should be possible to switch between sets if the task demands this (this is not the case in this task, but it is in Salvucci's [2005, this issue] driving task). The fifth set, select key, finds and presses a function key on the basis of a label. It is activated whenever a select key instruction is en-

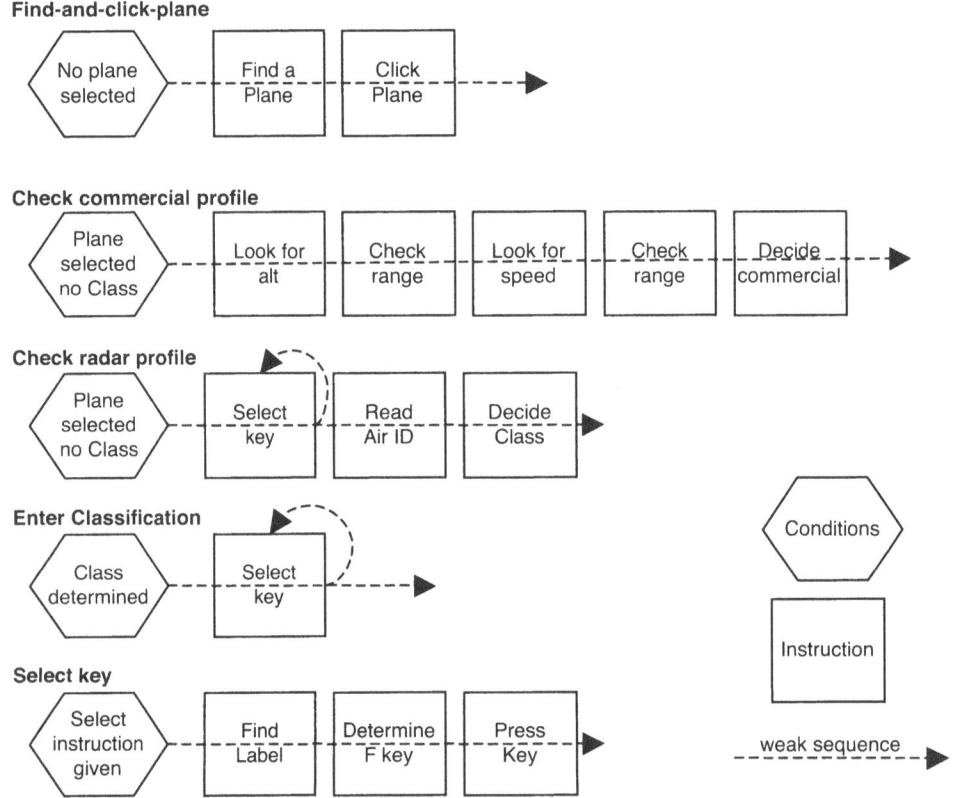

Fig. 10. Schematic representation of the instructions for CMU-ASP. Each row of boxes represents a rule set.

countered in one of the other sets, temporarily suspending that set. The instructions within a rule set are loosely ordered by activation. Instructions are retrieved by a rule and carried out by one or more subsequent rules. These subsequent rules also check the constraints on instructions. Production compilation combines instruction retrieval and interpretation rules into task-specific production rules that can fire out of sequence provided that their conditions are satisfied.

In the experiment reported by Anderson et al. (2004), participants had to solve twenty 6-min CMU-ASP scenarios spread over two sessions of 10 scenarios each. Fig. 11 shows the global behavioral outcomes of the experiment and the fit with the model. Fig. 11a shows the number of planes successfully classified in each scenario, whereas Fig. 11b shows the average time per scenario to perform one of three subtasks. *Select* is the elapsed time until a plane is clicked and corresponds to the *find-and-click-plane* rule set. *Classify* is the time between the mouse-click and the first keystroke of entering a classification (the "Track manager" keystroke) and covers two subtasks that cannot be separated easily: *Check-commercial-profile* and *Check-radar-profile*. Finally, *Enter* covers the time between the first and the last classification keystroke, that is, the sequence of eight keystrokes needed to enter the classification into the system. Although the general match between model and data is good, it is by no means better than that for Anderson et al.'s (2004) model. To see the difference, we have to examine the details of the new model's learning.

3.4. Finding and clicking a plane

The first step of identifying a plane is selecting a plane on the radar screen and clicking it. In the model these two instructions make up the first instruction set, which is applicable whenever there is no plane selected. The first instruction is to find a visual object of type *square-plane* on the screen and memorize it (*find-object*), and the second instruction is to click on the memorized object (*click-object*). Neither of these two instructions can be carried out in one step, in that both involve three steps. To do a *find-object*, first the location of an object of the appropriate type must be found on the screen. The next step is to initiate an eye movement to the object and to attend it. The third step is to process the information of the object and see whether it is a candidate for selection (when the track has been processed before, it is not a candidate). The *click-object* instruction requires first moving the hand to the mouse (if it is not already there), then moving the mouse onto the object, and finally clicking the mouse. Doing all these six steps (three from *find-object* and three from *click-object*) in sequence is inefficient, because the first three are perceptual actions and the last three are manual actions. Treating the execution of the two instructions as a dual-task situation (with restrictions) can lead to more efficient behavior. Initially the model will carry out all the steps in sequence by a retrieve-instruction and interpret-instruction cycle:

Retrieve-instruction
IF   the goal is to do a rule set (goal buffer)
THEN   retrieve an instruction from that rule set (retrieval buffer)

Implement-attend-location
IF   the goal is to do a rule set (goal buffer)
AND   the instruction is to attend a visual stimulus of a certain type (retrieval buffer)
AND   the visual module has found an object of that type (visual buffer)
THEN   move the eyes to that object and attend it (visual buffer)

Number of planes classified

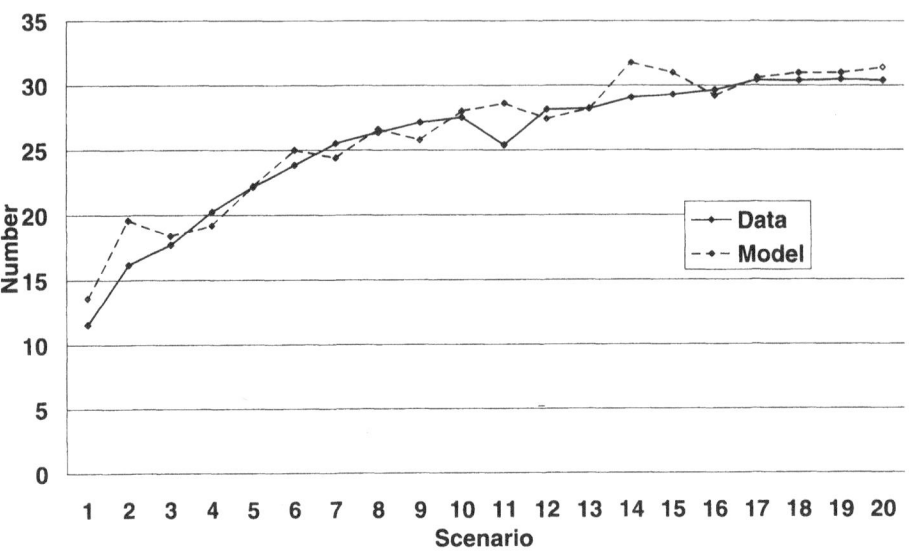

Fig. 11a. Results of the Anderson et al. (2004) experiment and the model: number of planes successfully classified per scenario.

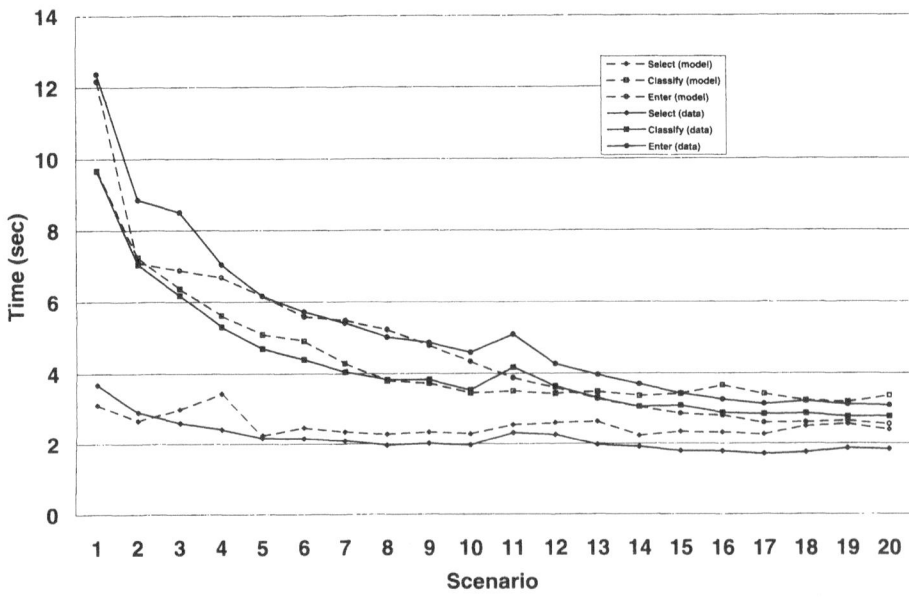

Fig. 11b. Results of the Anderson et al. (2004) experiment and the model: average time to complete a unit task per scenario.

The first of these rules will retrieve any instruction from this set. Activation between instructions will initially give *find-object* the highest activation, but, once *find-object* has been completed, *click-object* will have the highest activation. Although this activation scheme initially produces the same behavior as pure top-down models, the behavior changes once production rules are learned. The previously mentioned second rule is an example of a possible implementation rule. On the basis of the previous two rules in combination with the instruction to attend a square object, learning produces the rule:

Learned-attend-rule
IF the goal is to do rule set find-and-click-plane (goal buffer)
AND we have found a location on the screen of type square-object (visual buffer)
THEN move the eyes to the location and attend it (visual buffer)

Similarly, the new model learns a rule to move the hand to the mouse:

Learned-hand-to-mouse-rule
IF the goal is to do rule set find-and-click-plane (goal buffer)
AND the hand is not on the mouse (manual buffer)
THEN move the hand to the mouse (manual buffer)

A property of both rules is that they can activate at any moment when the *find-and-hook-plane* rule set is active and a location of the right type is found, or when the hand is not on the mouse. Although performance was initially constrained by the sequenced retrieval of instructions, they now depend only on the constraints in the rule and this control state. This lets the model treat the two instructions as dual tasks. For example, the *learned-hand-to-mouse-rule* can fire as soon as the *find-and-hook-plane* rule set is activated because it does not need to know the object to be clicked. This has the advantage that the hand can be moved to the mouse right at the beginning of the *find-and-hook-plane* rule set, instead of waiting until a plane has been found on the screen. A second consequence is that perceptual processes can continue while the manual processes are busy. During the execution of the manual processes, other planes can be inspected. This is very useful for this task because there are certain criteria ranking the planes, and when a better plane is found it can be selected instead of the first choice.

Fig. 12 shows an example model trace of a novice. The structure of behavior is similar to that shown in the novice behavior in Fig. 3: An instruction is retrieved by a rule, it takes some time to retrieve the instruction, and then the instruction is carried out. Sometimes the execution of the instruction has to wait until the module in question is done with the previous action, which is most notable in the case of the motor actions. A difference with the dual-task model is that the same instruction is retrieved a number of times in a row, once for every step in the instruction. The figure also shows that there is no interleaving between the two instructions yet (represented by thick-bordered and thin-bordered boxes for, respectively, the *find-object* and *click-object* instructions).

Production compilation produces rules that can fire out of sequence, and it allows for an efficient interleaving of perceptual and motor actions (Fig. 13). There is no longer a need for retrieval from declarative memory, because all the instructions have been incorporated in the rules. In this example, perfect dual tasking of the two instructions is achieved. A single rule ini-

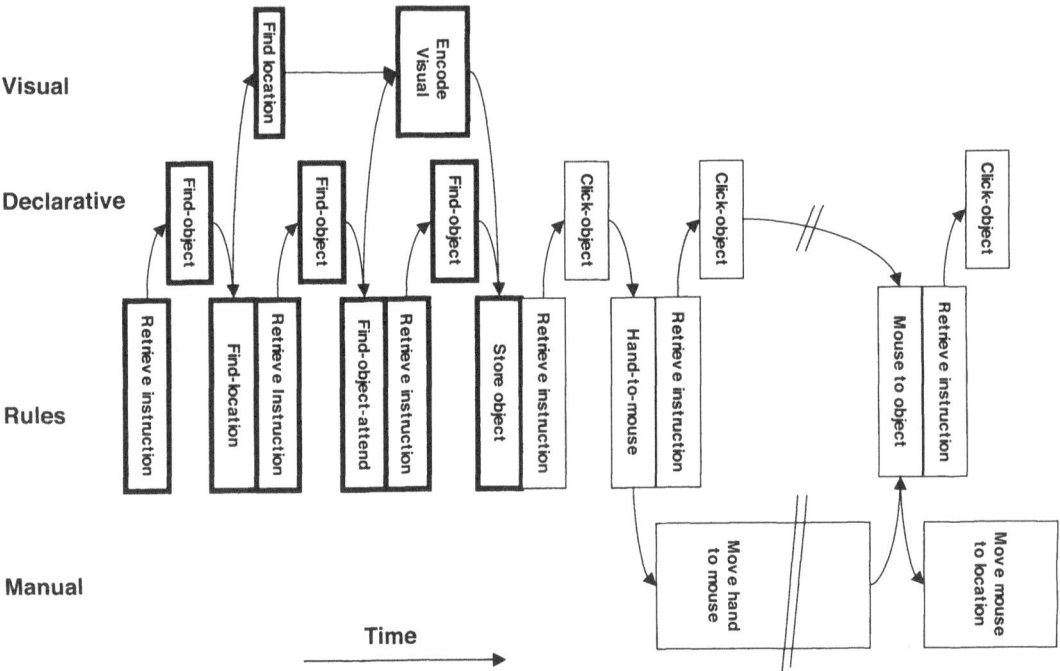

Fig. 12. Partial trace of novice model behavior on selecting a plane in the CMU-ASP task. Steps related to *find-object* are represented by thick-bordered boxes, whereas steps related to *click-object* are thin-bordered boxes.

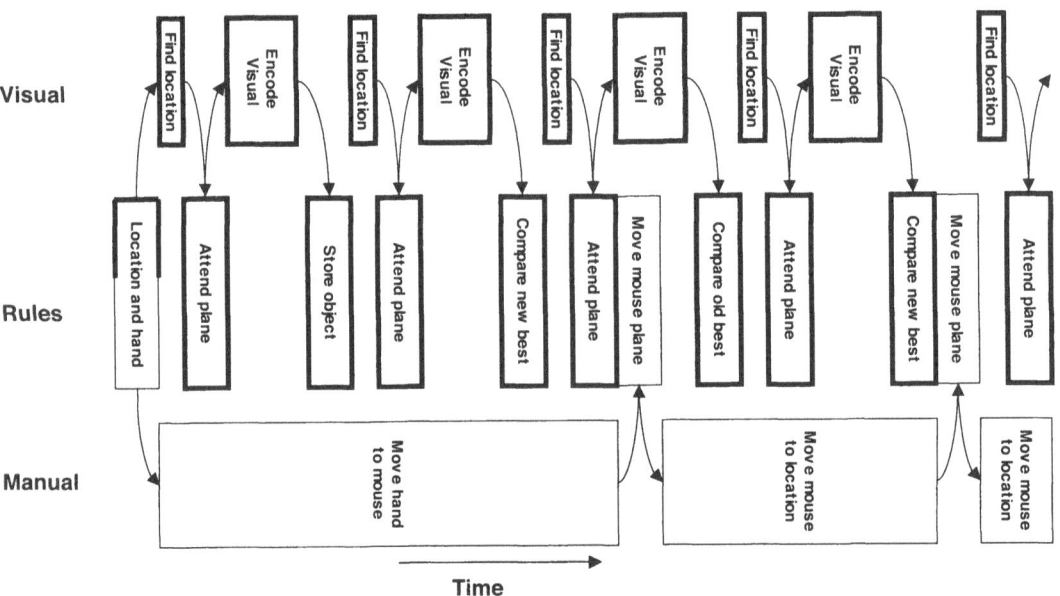

Fig. 13. Partial trace of expert behavior on selecting a plane.

tiates the movement of the hand to the mouse and the perceptual action to find a plane. During the hand movement a second plane is attended to and judged as better than the first plane. Halfway through attending to the third plane the hand movement finishes, enabling a rule that moves the mouse pointer to the best plane up to that moment. However, a fourth plane inspected during this mouse move is better than the plane to which the mouse is moving. Because this destination of the mouse movement is now no longer the target, a second mouse movement is initiated at soon as the first movement is completed.

This example shows that the same principles that produced parallelization in the Schumacher task (Schumacher et al., 2001) produce parallelization of instructions in CMU-ASP. The difference is that, in the CMU-ASP task, learning not only gives a (slight) improvement in speed, but also produces an improvement in the quality of the selection. Fig. 14 shows that this is true for the participants as well as the model, depicting the average point value of the first five planes selected by the model (based on the average of 12 model runs) and by the participants (based on 16 participants). Only the first five planes are used for the analysis, because early in the scenario the planes with the highest values stand out, whereas later in the scenario differences in point values become negligible. Both the model and the participants show a steady increase in selection quality. Initially (during Scenarios 1 and 2) the model does not compare planes, behaving as in Fig. 12, essentially choosing planes randomly. Only later does it compare planes, and it becomes gradually better at it. The participant behavior follows

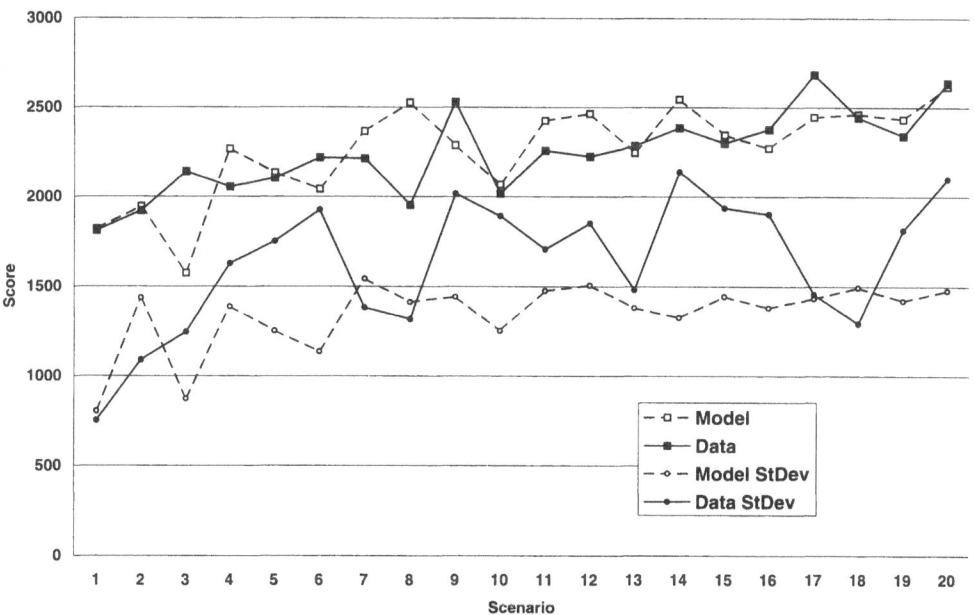

Fig. 14. Average en standard deviation of the point scores for the first five planes.

this trend, both in average scores and variance in the scores, suggesting they also move from random selection to a selection based on comparisons.

3.5. Learning the location of function keys

The *select* rule set contains instructions to press a function key given a label. It first attempts to find the label on the bottom of the screen (*find-label*), maps the location onto the function key (*determine-F-key*), and then presses that key (*press-key*). To explain participants' eventual speed, the model has to learn to press a function key given a label without first looking at the key mappings at the bottom of the screen. The model learns this by integrating several instruction steps, which lead to another case of dual tasking, but with an interesting twist. The two steps that overlap are the *find-label* and *determine-F-key* instructions. *Find-label* will try to find a certain text label (such as "Track Manager") on the screen by simultaneously scanning the display and trying to recall the location. If recall is successful, it will place the location of the recalled position in the visual buffer, directing the visual module to attend that location. The rule learned on the basis of that recall illustrates this:

Learned-Find-Track-Manager-Rule
IF   the goal is to select a key for label "Track Manager"
THEN   set the location in the visual buffer to coordinates (472,715) and request an
          attention shift to that location

Normally the model would wait until the attention shift is done and proceed from there. However, the next instruction, *determine-F-key,* only needs the location of the label to determine the right function key. This means that the key can be determined before the attention shift is done. The rule learned from retrieving the instruction is

Learned-Determine-F6-Rule
IF   the goal is to select a key
AND   the location in the visual buffer has x-coordinate 472
THEN   set the key to be pressed to F6

The *determine-F-key* instruction needs information that is available halfway through the *find-label* instruction. That means that at some point the rest of the steps to carry out *find-label* are obsolete and can be skipped. The whole process eventually only invokes two production rules, but this takes the model considerable time to learn: It must first compile rules based on the instructions, then it has to incorporate retrieved facts (locations of the labels) into these rules, and only then can it attempt to parallelize them, but they also have to win the competition with the already existing rules in terms of utility. A model with explicit states to sequence the instructions would need a planning system to determine that some steps are no longer needed, but by not having these control states the model discovers this by itself.

3.6. Learning to optimize hand movements

To press one of the function keys, either the left or the right hand can be used. The model has rules to use either the left or the right hand, but in practice the left hand is always on the left side

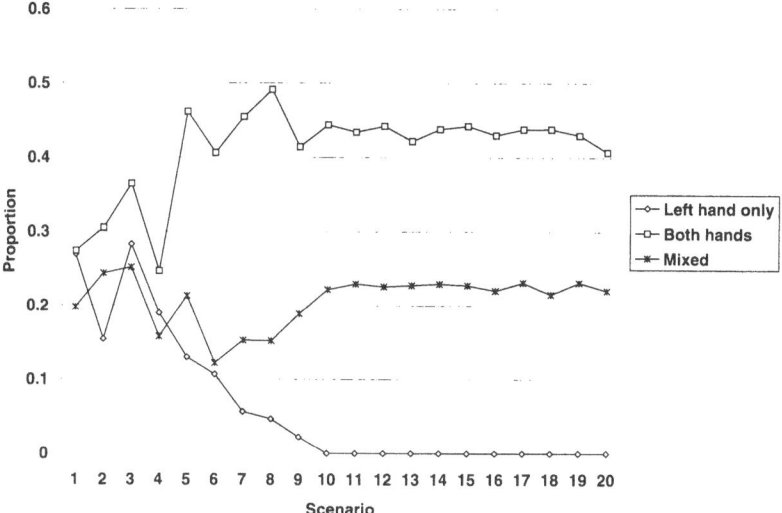

Fig. 15. Proportion of right-hand use for function keys. The function keys on the left side of the keyboard are also included, which are always pressed with the left hand.

of the keyboard, so is always the best choice for function keys on the left side of the keyboard. For the function keys on the right side of the keyboard the model can only use the right hand if the hand is on the keyboard and not on the mouse, otherwise it has to move the hand to the keyboard first. This allows for several different hand-movement styles. The model develops different styles in different model runs, as is depicted in Fig. 15. The figure shows how preference for using the right hand develops over the scenarios. The top curve in the graph comes from a case in which the model prefers to type with both hands, moving the hand back to the keyboard early in the classification process. The bottom curve comes from another run in which the model prefers to do all keying with the left hand and leaves the hand on the mouse. The middle curve is a compromise between the two: The model does the first few keystrokes with the left hand while the right hand is not yet back from the mouse. When the right hand has returned from the mouse, the remaining keystrokes are done with the right hand.

The model develops several styles because none of them are truly superior to the others. Initially the choice of style is random, but the first style to be proceduralized then has an edge over the styles that are not yet proceduralized, making that style even more attractive. In this way the model can end up with a strategy that is optimal only because it is the best practiced.

3.7. Conclusions

In the introduction we noted that many criticisms of rule-based systems focus on two common assumptions: the assumption that instructions are translated into rules right away and the assumption of a strong hierarchical task representation. The two models presented in this article do not rely on these assumptions and yet show a good correspondence with the empirical data. The instruction assumption is replaced by an initial representation of the task that has the form of declarative instructions. By interpreting these instructions, task-specific production rules are gradually learned. The strong hierarchical task representation assumption has been replaced by a weak two-level hierarchy guided by the minimal control principle, which states that models should use as few control states as possible.

The two models that we have constructed on the basis of these principles can learn their own rules and interact flexibly with both the external world and their own internal resources. The model of the Schumacher et al. (2001) experiments was able to achieve perfect interleaving of the two tasks although relying on only one control state. In the model of CMU-ASP, there is a mix between top-down and bottom-up influences: Instructions that are organized in rule sets are cued by both internal variables and external events. Initially each instruction within a rule set can be considered as a separate control state, but as production compilation creates rules that skip over instructions, eventually each rule set has just one control state. The collapse of control states within a rule set lets the model parallelize steps when possible and also lets new strategies emerge by spontaneous recombination of steps.

Kieras et al. (2000) assumed that skill acquisition goes through five stages with an emphasis on scheduling resources. In contrast to this focus, there are hardly any resource conflicts in the models discussed here. In some cases, conflicts are handled by the serial nature of the architecture: ACT–R processes requests for a resource on a first-come, first-served basis. If this leads to a conflict, it bases its choice on the utility of the competing rules. Another issue that Kieras et al. raised is the transfer of control between different tasks. In the dual-task experiments discussed here, the two tasks are very simple and presented at the same time, so they can be considered as one task. The CMU-ASP model does not involve multiple tasks, just subsequent instructions that are interleaved. When the two tasks are learned separately and each requires its own context, some explicit goal switch is needed (see Salvucci, 2005, this issue, for an example).

Although we addressed the issue of parallel versus serial cognition only in passing, the learning perspective casts a new light on the issue as well. ACT–R assumes that cognition is serial. To achieve true parallelism, very specialized production rules must be learned that address multiple tasks at the same time. Although this is feasible in situations such as the Schumacher et al. (2001) task, it only works in general in overly practiced combinations of tasks, or in cases where interleaving can be achieved easily. In EPIC, on the other hand, parallelism requires a scheduler, and the system must learn an optimal scheduler for the task. In the case of novice combinations of tasks, a general scheduler locks out one or more tasks, effectively enforcing seriality. The issue is therefore transformed into a contrast between ACT–R, which hypothesizes serial processing that must develop parallelism, and EPIC, which hypothesizes that parallel processing is blocked for new tasks, but that this blockage is gradually released. As long as EPIC has no explicit theory on how this blockage is removed, it is hard to assess whether the two theories produce equivalent predictions or not.

It is not clear to what extent our model results can also be achieved by the SOAR architecture (Newell, 1990). In SOAR, operator selection heavily depends on this state, but it is conceivable that in the EPIC-SOAR combination (Chong & Laird, 1997) this dependence could be diminished. Nevertheless, the general skill acquisition theory behind SOAR (Newell & Rosenbloom, 1981) seems to point in the direction of learning massive numbers of highly specialized rules for any combination of control state and input, instead of an attempt to reduce the number of possible control states.

Models of learning new tasks are not interesting only for studies of learning. They also constrain the space of possible expert models. Although the various modeling paradigms for expert models are capable of accurately describing performance, their formalisms allow for many models that do not correspond to reality. Learning models can show what is possible and what is not, and they may even exhibit surprising emerging behavior that the modelers never expected. The addition of the minimal control principle can constrain models even further: It guides model construction to choose the model with the fewest control states. Within ACT–R, the number of control states is easily quantified as the number of possible values of the goal state buffer.

We should note that the approach discussed here is currently limited to situations with a clear set of instructions. Other learning situations, such as learning from demonstration, analogy, and examples, are not yet captured by this framework. Another limitation of this work is that the modeler determines the way the whole task is partitioned into rule sets. Although each rule set must correspond to a single control state, offering some constraint, it would be more desirable if the model could determine its own rule-set boundaries. In future work we will explore how ACT–R models can generate their own instructions on the basis of examples and general knowledge and on partition rule sets when the need for additional control states arises.

Related to this issue is the fact that successful interleaving of tasks is only possible if the model knows the structure of the constraints between the instructions. In the CMU-ASP task, these constraints are fairly trivial: There is no reason why one could not move a hand to the mouse while looking at the screen, or typing a function key while not looking at its label. However, in other situations where instructions must be followed, the constraints can be hidden, especially if the instructions are written out as an explicit recipe, but the function of individual steps is unspecified. If a learner is confronted with such a situation, he or she can only proceduralize the sequence of steps, which will lead to faster execution, but not to increased flexibility. Examples of this problem can be found in programming video recorders (e.g., Gray, 2000) and operating flight management systems in modern airplanes (e.g., Sherry & Polson, 1999). In other words, poor instruction can lead to a situation in which a person has more control states than necessary.

Notes

1. This currently has no associated module, but see Salvucci (2005) for a proposal.

Acknowledgments

This research was supported by ONR grant N00014–04–1–0173 and NASA grant NCC2–1226.

I would like to thank John Anderson, Dan Bothell and Scott Douglass for providing me with the CMU-ASP data, and John Anderson, Stefani Nellen, Dario Salvucci, Pat Langley and an anonymous reviewer for their comments on earlier versions of the manuscript.

References

Ackerman, P. L. (1988). Determinants of individual differences during skill acquisition: Cognitive abilities and information processing. *Journal of Experimental Psychology: General, 117,* 288–318.

Anderson, J. R. (1982). Acquisition of cognitive skill. *Psychological Review, 89,* 369–406.

Anderson, J. R. (2005). Human symbol manipulation within an integrated cognitive architecture. *Cognitive Science, 29,* 313–341.

Anderson, J. R., Bothell, D., Byrne, M. D., Douglass, S., Lebiere, C., & Qin, Y. (2004). An integrated theory of mind. *Psychological Review, 111,* 1036–1060.

Anderson, J. R., Fincham, J. M., & Douglass, S. (1997). The role of examples and rules in the acquisition of a cognitive skill. *Journal of Experimental Psychology: Learning, Memory and Cognition, 23,* 932–945.

Anderson, J. R., Taatgen, N. A., & Byrne, M. D. (in press). Learning to achieve perfect time sharing: Architectural implications of Hazeltine, Teague, & Ivry (2002). *Journal of Experimental Psychology: Human Perception and Performance.*

Byrne, M. D., & Anderson, J. R. (2001a). ACT–R/PM and menu selection: Applying a cognitive architecture to HCI. *International Journal of Human–Computer Studies, 55,* 41–84.

Byrne, M. D., & Anderson, J. R. (2001b). Serial modules in parallel: The psychological refractory period and perfect time-sharing. *Psychological Review, 108,* 847–869.

Card, S. K., Moran, T. P., & Newell, A. (1983). *The psychology of human–computer interaction.* Hillsdale, NJ: Lawrence Erlbaum Associates, Inc.

Chong, R. S., & Laird, J. E. (1997). Identifying dual-task executive process knowledge using EPIC-Soar. In M. G. Shafto & P. Langley (Eds.), *Proceedings of the Nineteenth Annual Conference of the Cognitive Science Society* (pp. 107–112). Hillsdale, NJ: Lawrence Erlbaum Associates, Inc.

Dreyfus, H. L., & Dreyfus, S. E. (1986). *Mind over machine: The power of human intuition and expertise in the era of the computer.* New York: Free Press.

Fitts, P. M. (1964). Perceptual-motor skill learning. In A.W. Melton (Ed.), *Categories of human learning.* New York: Academic.

Freed, M., Matessa, M., Remington, R., & Vera, A. (2003). How Apex automates CPM-GOMS. In F. Detje, D. Dörner, & H. Schaub (Eds.), *Proceedings of the Fifth International Conference on Cognitive Modeling* (pp. 93–98). Bamberg, Germany: Universitats-Verlag.

Gray, W. D. (2000). The nature and processing of errors in interactive behavior. *Cognitive Science, 24,* 205–248.

Gray, W. D., John, B. E., & Atwood, M. E. (1993). Project Ernestine: Validating a GOMS analysis for predicting and explaining real-world performance. *Human–Computer Interaction, 8,* 237–309.

Hazeltine, E., Teague, D., & Ivry, R. B. (2002). Simultaneous dual-task performance reveals parallel response selection after practice. *Journal of Experimental Psychology: Human Perception and Performance 28,* 527–545.

Hodge, K. A., Rothrock, L., Kirlik, A. C., Walker, N., Fisk, A. D., Phipps, D. A., et al. (1995). Trainings for tactical decision making under stress: Towards automatization of component skills. (HAPL-9501). Atlanta: Georgia Institute of Technology, School of Psychology, Human Attention and Performance Laboratory.

Holland, J. H. (1986). Escaping brittleness: The possibilities of general purpose machine learning algorithms applied to parallel rule-based systems. In R. S. Michalski, J. G. Carbonell, & T. M. Mitchell (Eds.), *Machine learning II* (pp. 593–623). Los Altos, CA: Kaufmann.

Hornof, A. J., & Kieras, D. E. (1997). Cognitive modeling reveals menu search is both random and systematic. *Proceedings of CHI-97* (pp. 107–114). New York: Association for Computing Machinery.

Kieras, D. E., Meyer, D. E., Ballas, J. A., & Lauber, E. J. (2000). Modern computational perspectives on executive mental processes and cognitive control: Where to from here? In S. Monsell & J. Driver (Eds.), *Control of cognitive processes: Attention and performance XVIII* (pp. 681–712). Cambridge, MA: MIT Press.

Logan, G. D. (1988). Toward an instance theory of automatization. *Psychological Review, 95,* 492–527.

Logan, G. D. (1992). Attention and preattention in theories of automaticity. *American Journal of Psychology, 105,* 317–339.

Meyer, D. E., & Kieras, D. E. (1997). A computational theory of executive cognitive processes and multiple-task performance: Part 1. Basic mechanisms. *Psychological Review, 104,* 2–65.

Newell, A. (1990). *Unified theories of cognition.* Cambridge, MA: Harvard University Press.

Newell, A., & Rosenbloom. P. S. (1981). Mechanisms of skill acquisition and the law of practice. In J. R. Anderson (Ed.), *Cognitive skills and their acquisition* (pp. 1–55). Hillsdale, NJ: Lawrence Erlbaum Associates, Inc.

Pashler, H. (1994). Dual-task interference in simple tasks: Data and theory. *Psychological Bulletin, 116,* 220–244.

Rumelhart, D. E., & McClelland, J. (1986). *Parallel distributed programming* (Vol. 1–2). Cambridge, MA: MIT Press.

Salvucci, D. D. (2005). A multitasking general executive for compound continuous tasks. *Cognitive Science, 29,* 457–492.

Schumacher, E. H., Seymour, T. L., Glass, J. M., Fencsik, D. E., Lauber, E. J., Kieras, D. E., et al. (2001). Virtually perfect time sharing in dual-task performance: Uncorking the central cognitive bottleneck. *Psychological Science, 12,* 101–108.

Sherry, L., & Polson, P. G. (1999). Shared models of flight management system vertical guidance. *International Journal of Aviation Psychology, 9,* 139–153.

Taatgen, N. A., & Anderson, J. R. (2002). Why do children learn to say "broke"? A model of learning the past tense without feedback. *Cognition, 86,* 123–155.

Taatgen, N. A., & Lee, F. J. (2003). Production compilation: A simple mechanism to model complex skill acquisition. *Human Factors, 45,* 61–76.

Cognitive Science 29 (2005) 457–492

A Multitasking General Executive
for Compound Continuous Tasks

Dario D. Salvucci

Department of Computer Science, Drexel University

Received 3 May 2004; received in revised form 31 July 2004; accepted 9 November 2004

Abstract

As cognitive architectures move to account for increasingly complex real-world tasks, one of the most pressing challenges involves understanding and modeling human multitasking. Although a number of existing models now perform multitasking in real-world scenarios, these models typically employ *customized executives* that schedule tasks for the particular domain but do not generalize easily to other domains. This article outlines a *general executive* for the Adaptive Control of Thought–Rational (ACT–R) cognitive architecture that, given independent models of individual tasks, schedules and interleaves the models' behavior into integrated multitasking behavior. To demonstrate the power of the proposed approach, the article describes an application to the domain of driving, showing how the general executive can interleave component subtasks of the driving task (namely, control and monitoring) and interleave driving with in-vehicle secondary tasks (radio tuning and phone dialing).

Keywords: Multitasking, Cognitive architectures, ACT–R, Driving

1. Introduction

As theories of cognition have matured over the last decade, they have increasingly broadened in scope beyond simple laboratory tasks to a wide variety of complex, dynamic task domains. In particular, the instantiation of cognitive theories as unified computational cognitive architectures (e.g., Anderson et al., 2004; Just, Carpenter, & Varma, 1999; Meyer & Kieras, 1997; Newell, 1990) has facilitated application and validation of these theories in real-world domains; for instance, researchers have used cognitive models to better understand human behavior in such domains as air traffic control (e.g., Taatgen & Lee, 2003), human–computer interaction (e.g., Kieras, Wood, & Meyer, 1997; Ritter, Baxter, Jones, & Young, 2000), game playing (e.g., Lebiere, Gray, Salvucci, & West, 2003), even aircraft piloting (Jones et al., 1999). Such modeling efforts for complex tasks have brought to light many new challenges for

---

Requests for reprints should be sent to Dario Salvucci, Department of Computer Science, Drexel University, 3141 Chestnut Street, Philadelphia, PA 19104. E-mail: salvucci@cs.drexel.edu

cognitive architectures, and one of the most critical yet elusive challenges has been the modeling of general human multitasking—how people integrate and perform multiple tasks in the context of a larger complex task.

A number of models developed in cognitive architectures have now emerged that, either explicitly or implicitly, account for aspects of human performance while multitasking. Table 1 shows a sampling of multitasking models classified into four broad categories as identified by Kieras, Meyer, Ballas, and Lauber (2000). Models of *discrete successive tasks* examine performance in alternating trials (or series of trials) of simple choice-reaction tasks, accounting for

Table 1
Examples of multitasking models developed in a cognitive architecture

Domain	Architecture(s)	Reference
Discrete Successive Tasks		
Alternating choice	ACT–R	Altmann & Gray, 2000
Alternating choice	EPIC	Kieras et al., 2000
Alternating choice	ACT–R	Sohn & Anderson, 2001
Discrete Concurrent Tasks		
Dual choice	EPIC	Meyer & Kieras, 1997
Dual choice	ACT–R	Byrne & Anderson, 2001
Dual choice	ACT–R	Anderson, Taatgen, & Byrne, in press
Elementary Continuous Tasks		
Tracking and choice	EPIC	Kieras & Meyer, 1997
Tracking and choice	EPIC–SOAR	Chong, 1998
Tracking and choice	SOAR, EPIC	Lallement & John, 1998
Compound Continuous Tasks		
Air traffic control (KA-ATC)	ACT–R	Taatgen & Lee, 2003
Air traffic control (AMBR)	ACT–R, D-COG, EPIC–SOAR, iGen	Gluck & Pew, in press
Aircraft maneuvering	ACT–R	Gluck, Ball, Krusmark, Rodgers, & Purtee, 2003
Aircraft piloting (TacAir–SOAR)	SOAR	Jones et al., 1999
Aircraft piloting	CI/ADAPT	Doane & Sohn, 2000
Aircraft taxiing	ACT–R	Byrne & Kirlik, 2005
Driving	SOAR	Aasman, 1995
Driving	QN-MHP	Tsimhoni & Liu, 2003
Driving and phone dialing	ACT–R	Salvucci, 2001
Dynamic systems	ACT–R	Schoppek, 2002
Game playing (Quake)	SOAR	Laird & Duchi, 2000
Game playing (Unreal Tournament)	ACT–R	Best & Lebiere, 2003
Radar operation (Argus Prime)	ACT–R	Gray & Schoelles, 2003
Shooting and mathematical comprehension	ACT–R, IMPRINT	Kelley & Scribner, 2003
Tactical decision making and instruction following	ACT–R	Fu et al., 2004
Tracking and decision making	EPIC	Kieras & Meyer, 1997

Note. ACT–R = Adaptive Control of Thought–Rational; EPIC = Executive-Process Interactive Control; D–COG = Distributed Cognition; CI/ADAPT = Construction-Integration/ADAPT; QN–MHP = Queuing Network—Model Human Processor.

the temporal costs of switching from one task to another. Models of *discrete concurrent tasks* analyze performance in concurrent choice-reaction tasks typically offset by a short delay, accounting for "psychological refractory period" effects of dual-task interference. Models of *elementary continuous tasks* address behavior when integrating a continuous task with occasional short discrete tasks—for instance, performing a manual tracking task (e.g., keeping a cursor on target) while occasionally responding to a choice-reaction task. These first three categories all include at least one discrete choice-reaction task that lasts at most a few seconds. In contrast, models of *compound continuous tasks* account for behavior in two or more simultaneous tasks, each of which is an ongoing continuous process or at least takes enough time to require interleaving with other tasks. Compound continuous tasks abound in real-world domains, particularly in the space of complex dynamic tasks that have become a centerpiece of cognitive modeling, and serve as the primary focus of this article.

Models of compound continuous tasks, such as those highlighted in Table 1, have begun to elucidate important aspects of management and coordination of multiple component tasks. For instance, Jones et al.'s (1999) fighter piloting model can identify new aircraft on radar while flying to intercept other aircraft; Lee and Anderson's (2000) air traffic control model can accept and land planes while monitoring landing conditions such as weather; and Fu et al.'s (2004) model can listen to and interpret "over-the-shoulder" instructions while performing a tactical decision-making task. Typically, analysis and validation of the model's behavior focuses on aggregate performance in the task, much of which can be affected by multitasking requirements; for instance, Kieras and Meyer's (1997) model of manual tracking demonstrated larger tracking errors in the presence of more difficult (i.e., more visually eccentric) choice tasks, and my own model of driving and phone dialing (Salvucci, 2001b) exhibited adverse effects of dialing on overall steering performance. In addition, a few models have been analyzed specifically for how and when participants switch between component tasks; for example, Gray and Schoelles's (2003) radar-operator model demonstrated how even models that accurately capture aggregate performance can miss important aspects of task-switching-specific measures, such as how often people switch at unit-task boundaries. All these efforts contribute to a broader understanding of multitasking through study of both overall measures of task performance and particular measures of multitasking performance.

Even with this wide diversity of models and domains, current models of compound continuous tasks share one limitation: the use of *customized executives* (Kieras et al., 2000), namely, multitasking control mechanisms that have been specialized and fine-tuned for a specific model or domain or both. Fig. 1 illustrates how a customized executive unifies two individual task models into an integrated model that performs both tasks. Given two models, A and B, the customized executive is designed to integrate these two specific models, resulting in a specialized executive for the integrated model A + B. Such an executive generally does not transfer to new models C and D, thus requiring a new customized executive for the integrated model C + D. In addition, model integration with a customized executive often requires modification to the component models themselves (illustrated by A′, B′, etc., in the figure), because one model typically explicitly passes control to the other model and must thus be aware of this model (and likely some of the declarative structures associated with it). The lack of transfer and need for component-model changes means that every domain requires a new approach and new models

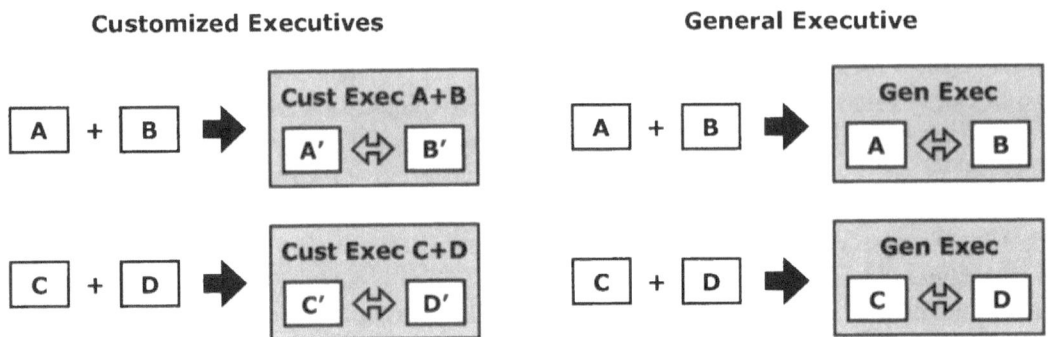

Fig. 1. Conceptual overview of customized executives and a general executive.

of multitasking, hampering both theoretical consistency among models of different domains as well as practical development through model reuse and transfer.

In the search for a general theory and model of human multitasking, we strive to move beyond customized executives to a unified *general executive*. As illustrated in Fig. 1, a general executive integrates separate models of task behavior and predicts the behavior that results from combining these models into a more complex, multitasking integrated model. Ideally, the general executive applies across all domains, providing us with a domain-independent theory of multitasking and an associated computational model of how multitasking takes place. Also, the general executive requires no modifications to the component models: Just as people can learn individual skills independently and then integrate these skills into multitasking behavior, the general executive can take models developed or learned independently of one another and unify these models into a psychologically plausible integrated multitasking model. (Subsequent learning could then specialize the component models or the executive itself for better multitasking performance; for now, we ignore such potential learning, although we address several issues related to learning later in the article.) Thus, rather than a collection of customized executives each specialized to particular domains, the general executive posits a broader theory that helps to unify our understanding and modeling of human multitasking.

This article describes a first step toward this general theory of multitasking, namely a formulation of a general executive for compound continuous tasks framed in the context of the ACT–R (Adaptive Control of Thought–Rational) cognitive architecture (Anderson et al., 2004). The general executive centers on a temporally aware queuing mechanism that manages current task goals and posits that goal representations are guided in part by reasonable heuristics of multitasking behavior. The article then demonstrates an application of the general executive to the domain of driving—a very rich, complex, ubiquitous task that millions of people perform on a daily basis. This application illustrates how the general executive accounts for driver multitasking both in integrating component subtasks of driving itself—namely, control and monitoring—and in integrating driving with a secondary task, namely, tuning a radio and dialing a cellular phone. Although the driving domain does not itself suffice as a complete validation of the general executive, it does serve as an excellent representative of compound continuous tasks (such as those in Table 1) that require management and scheduling of multiple processes in a complex environment.

2. A multitasking general executive for the ACT–R cognitive architecture

The development of a general executive and its instantiation in the ACT–R architecture serve as both a theoretical and practical endeavor. The primary motivation for the general executive is a cognitively plausible theory of multitasking that, like all aspects of a cognitive architecture, obeys the limitations and constraints of human cognition and performance. A secondary goal of the general executive is the facilitation of model integration and prediction—that is, the integration of independent individual task models into unified models and subsequent prediction of the effects of one component task on the others. This section begins with a description of the guiding theoretical principles on which the proposed general executive is based. It then continues with the specification of the executive in the ACT–R architecture, previously outlined in a preliminary report (Salvucci, Kushleyeva, & Lee, 2004) and generalized here.

2.1. Guiding principles for the general executive

Although our understanding of a multitasking general executive is very much incomplete, empirical and modeling research in this and related areas have begun to flesh out guiding principles for a possible executive model. This effort incorporates three guiding principles that helped to shape the general executive proposed here, as outlined in the following sections.

2.1.1. The general executive is an architectural mechanism

One possible approach to modeling multitasking in a cognitive architecture views multitasking as a learned cognitive skill like any other, and thus the mechanism should be instantiated as a model that abides by all architectural constraints rather than as a specialized mechanism. Such an approach requires that the component models in the integrated model somehow pass execution from one to another; one model may serve as a type of executive and help schedule and manage tasks, or all models may act as sibling processes and share execution time through a predetermined scheme. In either case, this view has at least two problematic issues that make it less plausible for a multitasking general executive. First, because models are explicitly shifting execution to others, each model must be aware of the other processes and must have special rules for passing control to these models. If models are learned or developed independently of one another, it is difficult to imagine how such specialized rules would evolve. Second, a general executive must have some form of interruption to ensure that running models are given a fair share of time, especially because, again, independently learned models would not necessarily know to give up control in a fair manner. If the general executive operates at the level of all other acquired skills, it is not clear how such a model would be able to interrupt other component models and shift execution to other models.

An alternate approach, and the one followed here, views the general executive as a specialized mechanism implemented within the cognitive architecture. This approach more easily addresses the issues mentioned previously: An architectural general executive would not require component models to be aware of one another, and the executive could certainly be made to interrupt the component models (leaving the deeper issue of when and how to interrupt). In addition, this approach seems more in line with neuropsychological evidence that suggests that general executive processes are located in different regions of the brain than procedural rule-based processes: Whereas prefrontal brain regions have been associated with mainte-nance of goal state (Koechlin, Corrado, Peitrini, & Grafman, 2000; Smith & Jonides, 1999) and the dorsolateral prefrontal cortex with goal memory and planning (Fincham, Carter, van Veen, Stenger, & Anderson, 2002), the basal ganglia have been associated with procedural memory and rule-based processes (Anderson et al., 2004). Thus, the general executive pro-posed here follows this view of the general executive as an architectural mechanism.

2.1.2. The general executive is dependent on time

When given a decision about what active task may proceed, the general executive must strike a delicate balance in allocating a fair amount of time to each active task. One way to achieve this balance is by selecting tasks on the basis of urgency: As a task is "starved" for re-sources (i.e., not allowed to proceed) for some period of time, its urgency steadily rises, thus making it more likely to be selected to proceed. To complicate this straightforward view, tasks in fact differ with respect to their desired running time—that is, not all tasks may wish to pro-ceed immediately but rather can afford to wait some time before proceeding. As one example, in an air traffic control task where the wind direction changes every 30 sec, F. J. Lee (personal communication, April 19, 2005) found that participants were most likely to examine wind di-rection right around the 30-sec mark. As another example, Kushleyeva, Salvucci, and Lee (2005) showed that participants tended to switch from one visual search task to another at the halfway point of the full 30-sec task period to ensure they were not penalized on the other task. In both cases, people tended to perform the timed tasks not as soon as possible, but rather ap-proximately at the time they deemed appropriate, given the temporal characteristics of the task.

Some of the best examples of the temporal dependence of multitasking arise in the many studies of driver distraction. Wierwille (1993) summarizes results from several studies, such as one by Dingus, Antin, Hulse, & Wierwille (1989), examining glance durations for a wide array of secondary tasks, including checking fuel, changing the radio station, and setting the vehicle defroster. In this study, the total number of glances inside the vehicle ranged from approxi-mately one to seven glances, but the average duration per glance consistently fell between 0.6 and 1.6 sec. Thus, drivers exhibited an acute sense of how long they looked inside the vehicle, knowing they could safely look for a short time (approximately 1 sec) but that, as time passed, the urgency to return to driving steadily increased. Such temporal dependence appears in many real-world dynamic environments in which the passage of time corresponds with increasing uncertainty about the world. Thus, the general executive requires some formulation that can in-corporate different task urgencies as part of its task management and scheduling.

2.1.3. The general executive is sensitive to goal representations

In addition to temporal issues of when to switch tasks, we can also ask how task and goal representations may influence multitasking—that is, are there "strongly connected compo-nents" of task procedures such that people are less likely to switch within a component and more likely to switch between components? If so, are these strongly connected components tied to representations of the goal structure? Empirical studies of extreme dual tasking, such as psychological refractory period studies mentioned earlier, suggest that with practice people can interleave tasks at the proposed smallest unit of cognitive cycle time, namely 50 msec (see, e.g., Card, Moran, & Newell, 1983; Meyer & Kieras, 1997; Newell, 1990). However, under general conditions of integrating multiple task processes in compound continuous tasks, studies suggest that task switching can indeed be highly influenced by goal representations and the "strongly connected components" of task procedures within these representations. For instance, in a radar operator task, Gray and Schoelles (2003) found that task switching most likely occurred between "unit tasks"; switching also occurred within unit tasks, but with lower probability. As another example, later in this article we examine the case of phone dialing while driving, where the data suggest that drivers update steering between the (3- and 4-digit) chunks of a phone number. Thus, a general executive should be flexible enough to allow goal and other declarative representations to influence task interleaving.

The principle of switching at natural representational boundaries has already been followed explicitly or implicitly by some existing models of compound continuous tasks with customized executives. For instance, Gluck et al.'s (2003) aircraft maneuvering model establishes new control settings under two specific conditions—at the beginning of a trial, and whenever the assessment of a control instrument shows large deviations from desired values—and thus implicitly defines the most natural switching points for aircraft maneuvering. Moreover, some modeling architectures themselves incorporate mechanisms to provide "robustness against interruption" to discourage or prevent preemption at critical points in processing (e.g., APEX: Freed, 1998). Thus, specific models and architectures have already recognized and begun to address the need for influences of goal representations on multitasking, and the general executive proposed here attempts to incorporate these ideas into a domain-independent mechanism for the ACT–R architecture.

2.2. An ACT–R general executive

The previously mentioned principles do not specify a general executive in their own right, but instead provide guidance for specifying a computational mechanism and for instantiating the executive within a cognitive architecture. The following section describes one possible instantiation of these guiding principles in the ACT–R cognitive architecture, with the overall goal of specifying a straightforward mechanism that accounts for several aspects of multitasking observed in real-world complex tasks, particularly for compound continuous tasks and the driving domain in the next section. Although this article focuses on the ACT–R architecture as the context for the general executive, the guiding principles and even aspects of the ACT–R instantiation that follows should generalize well to other architectures and modeling frameworks.

2.2.1. ACT–R and multitasking

The ACT–R cognitive architecture (Anderson et al., 2004; see also Anderson, 2005, this issue) is a production system architecture with a declarative memory store for factual chunks and a procedural memory store for condition–action production rules. The architecture incorporates a number of modules that interact with cognition by passing information through *buffers*—for instance, a visual buffer that holds the result of the vision module's visual encoding, or a motor buffer that passes movements to the motor module. One such buffer is the *goal buffer*, which stores this goal (itself a declarative chunk) and guides the production system to work on its associated task. Multitasking thus has a very straightforward interpretation in the architecture as the switching of goals in the goal buffer, thus allocating a period of cognitive processing time to one goal or task and then, after some time, switching to another goal or task.

Interestingly, the current ACT–R does not have a full-fledged goal module associated with the goal buffer; instead, new goals are managed and set directly through the actions of production rules. However, the previous version of the architecture (ACT–R 4.0, described in Anderson & Lebiere, 1998) incorporated a goal stack onto which new goals could be "pushed" and from which completed goals could be "popped." In essence, this goal stack served as one instantiation of a possible goal module, implicitly positing that people maintain goals and subgoals in a hierarchical stack representation. The goal stack was abandoned in the subsequent architecture due primarily to the psychological implausibility of such a stack because of memory limitations (Altmann & Trafton, 2002). The general executive described here can be viewed as a proposal for a new goal module, taking on the responsibility of maintaining and scheduling multiple goals and offloading this responsibility from the individual task production rules.

2.2.2. Architectural module for the general executive

The proposed ACT–R general executive represents a new goal module that maintains and schedules a set of active goals. In the current ACT–R, production rules set new goals by creating a declarative goal chunk and storing this chunk in the goal buffer, effectively replacing the old goal with the new one—for instance (in pseudocode approximating ACT–R production rules):

Do-Next-Task
IF   the goal is to perform Task0
            and this task is complete
THEN   replace the current goal with a goal to perform Task1

In contrast, the proposed ACT–R general executive allows for addition or removal of goals to or from a set of active goals—for example,

Do-Multiple-Next-Tasks
IF   the goal is to perform Task0
            and this task is complete
THEN   remove the current goal
            and add new goals to perform Task1 and Task2

In adding both new goals to the active set, the goal module is now charged with managing both goals and overseeing their execution. This policy of addition and removal still allows for goal replacement by simply removing this goal and adding a new one; at the same time, the new policy provides the additional feature of allowing a model to have and manage multiple goals.

Even with multiple active goals, the ACT–R architecture still posits only one goal buffer and thus can only perform one goal at a time. This being the case, how does the module determine which active goal should proceed next? For this purpose, the proposed general executive employs a *goal queue* that operates on a first-come, first-served (FCFS) basis: Whenever rules add new goals, the goals are placed on the goal queue and "served" (i.e., allowed to proceed) in the order in which they arrive at the queue. Queues have been studied in detail in the applied context of computer and networking systems and the more theoretical context of general queuing theory (see, e.g., Nelson, 1995). However, queues have also been studied as psychologically based constructs for modeling visual information processing (Ellis, Goldberg, & Detweiler, 1996) and general behavior (Miller, 1993), along with conceptual relatives such as cascade models (e.g., McClelland, 1979). Of note, Liu (1996) provided a detailed discussion of queuing networks of elementary cognitive processes, including the assumption of FCFS single-channel processing. The general executive proposed here follows, albeit in a simplified way, in this tradition with its integration of a queue as a basic psychological mechanism.

To expand on the production rule example previously mentioned, Task1 and Task2 are added simultaneously, and thus let us assume that Task1 proceeds by random chance (to be clarified in a later section). Also, let us assume that these goals are each associated with a single production that simply iterates one instance of this goal after another for purposes of illustration:

Iterate-Task1

IF   the goal is to perform Task1
          and this task is complete
THEN   remove the current goal
          and add a new goal to perform another iteration of Task1

Iterate-Task2

IF   the goal is to perform Task2
          and this task is complete
THEN   remove the current goal
          and add a new goal to perform another iteration of Task2

When Task1 proceeds by firing the *Iterate-Task1* rule, the rule creates a new Task1 goal and adds this goal to the queue. At this point, the queue contains the older Task2 goal added in the first firing and this newer Task1 goal added at the second firing, and thus the executive chooses the older Task2 goal to proceed. When Task2 proceeds by firing the *Iterate-Task2* rule and creating a new Task2 goal, the active Task1 goal now takes precedence and proceeds. Thus, the goals alternate positions in the goal queue, resulting in alternating firings of *Iterate-Task1* and *Iterate-Task2*.

This example represents a simplistic case in which each goal is associated with a single production firing before creating a new goal; in general, a goal will proceed for several rule firings before creating another goal. For this general case, the general executive selects a new goal and alters the goal buffer only when a rule firing creates a new goal or removes this goal—that is, whenever a rule firing modifies the goal queue in any way. A consequence of this is that a goal can proceed unhindered so long as it does not create or remove goals (forcing us to further consider the granularity of goal representations, as discussed shortly). Consider the example illustrated in Fig. 2(a), which assumes that Task1 requires two rule firings for a total of 100 msec on every iteration, and Task2 requires three rule firings for a total of 150 msec on every iteration. Assuming Task1 proceeds first, the Task1 rules fire and complete the first iteration, with the final rule firing creating a second Task1 goal. At this point, the general executive interrupts Task1 and allows the first Task2 goal to proceed until completion. This Task2 goal creates a second Task2 goal, allowing the second Task1 goal to intercede. Thus, like the earlier example,

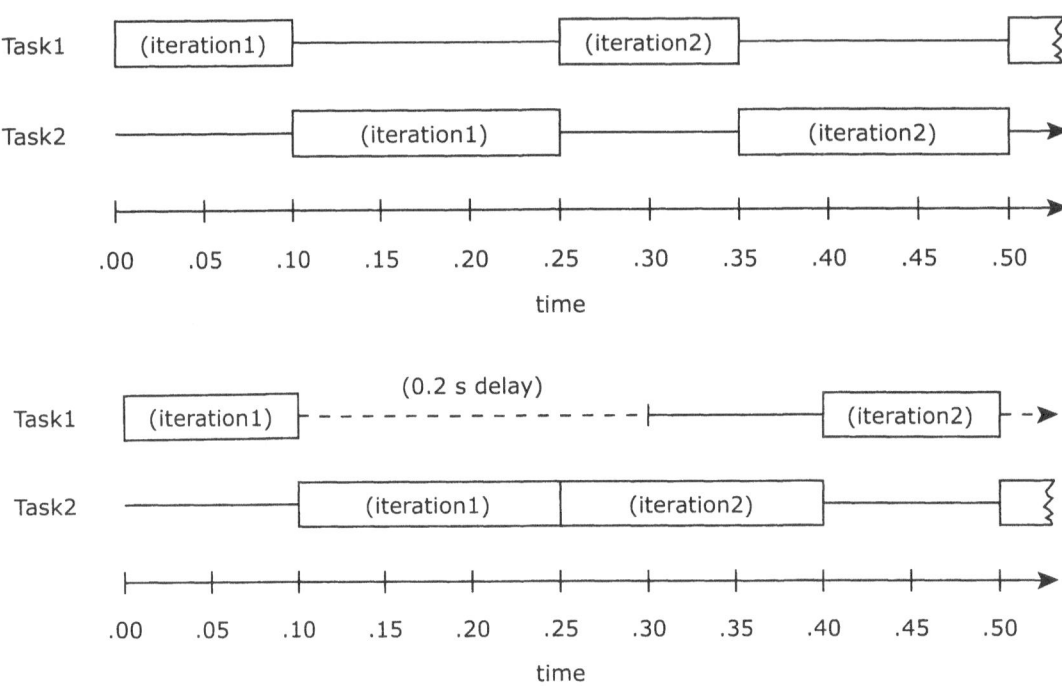

Fig. 2. Timeline for two tasks with (a) no delay on either task, and (b) a 0.2 sec delay for *Task1*. Each box indicates the time span during which the given goal would be executed.

the general executive alternates between goals for the two tasks, but only interrupts a task when it creates a new goal.

2.2.3. Incorporation of temporal dependence

The FCFS queue already incorporates one type of temporal dependence, namely, that goals are allowed to proceed in the temporal order in which they are created. However, as noted earlier, it is desirable that the general executive also addresses the fact that some tasks do not require immediate processing but rather wish to or can afford to be delayed for some period of time. To this end, the basic goal queue is augmented by allowing goals to specify a desired *delay* relative to the current time—for instance,

> *Iterate-Task1*
> *IF*   the goal is to perform Task1
>   and this task is complete
> *THEN*   remove the current goal
>   and add a new goal to perform another iteration of Task1
>   after a delay of 0.2 sec

The delay defaults to a value of 0 if not specified (i.e., wishes to run immediately), as in the *Iterate-Task2* production earlier, and also allows for a special *now* value that triggers immediate execution of this goal regardless of the state of the goal queue. For our purposes, we assume that the delay can be up to a few seconds into the future; prospective goals that are minutes, hours, or more in the future involve a number of other issues, such as retrieval failures, and are beyond the scope of this article.

To incorporate this temporal dependence into the goal queue, each goal in the queue is associated with a desired start time, specifically the goal's time of creation plus any specified delay. (Note that for goals with a default delay of 0, the desired start time is equal to creation time, indicating their desire to start immediately if possible.) When the production system adds or removes goals, and the goal module reevaluates the goal queue, the module chooses the goal with the earliest start time and allows this goal to proceed. In essence, this scheme implements a slightly modified FCFS queue: If all goals have a default delay of 0, the goal queue operates as a standard FCFS queue, but the queue also allows for nonzero delay times that reorder the queue and account for delayed goal execution. Equivalently, this scheme can be formulated as using urgency for scheduling: Each goal has an urgency defined as the negation of the remaining time before the goal wishes to run (i.e., *−delay*), and the general executive chooses the goal with the highest urgency as the next to proceed.

Fig. 2(b) illustrates the delay mechanism used in conjunction with our earlier example: the Task1 and Task2 rules remain the same except that the Task1 final rule specifies a delay of .20 sec before the next Task1 goal. As before, the first Task1 goal proceeds until completion at time .10 sec, then creates a new goal with a delay of .20 sec, thereby requesting a desired start time of .30 sec. The first Task2 goal intercedes as before, but this time on completion, the next Task2 goal has an earlier start time (.25 sec) than the next Task1 goal (.30 sec). Thus, the second Task2 goal proceeds until completion, at which point the second Task1 goal has waited .10 sec and is finally allowed to proceed.

Given the goal queue's reliance on timing information and the fact that people have an imperfect perception of time, the mechanism also includes systematic noise in the goal queue ordering to incorporate limited stochasticity to the process. Several aspects of the ACT–R architecture (e.g., chunk activation, production-rule conflict resolution) already incorporate noise, specifically a logistic distribution of noise with variance $s^2 = p^2 s^2 / 3$ driven by a noise parameter s. To maintain consistency with the architecture, the goal queue includes logistic noise around the start times of the individual goals, along with a parameter s_{gq} (gq for "goal queue") to specify the amount of this noise. Thus, the previous example indicates only one possible scheduling of goals and tasks, but this scheduling may change, given larger values of temporal noise in the process. Also, the noise results in essentially a random choice between goals with the same desired start times, such as goals created by the same production rule with the same delay time.

2.2.4. Heuristics for goal representations

As mentioned, one of the key issues with any multitasking general executive arises in the delicate balance between allowing a task to proceed versus preemptively interrupting the task to allow other tasks to proceed. The proposed general executive strikes this balance by potentially interrupting and switching tasks *between* goals without interrupting *within* goals. In essence, this strategy is motivated by the earlier discussion of the influences of goal representations on multitasking, finding "natural" breaking points in which to interleave other tasks. In theory, this strategy is sufficient for interleaving any number of ACT–R models and exhibiting multitasking behavior. In practice, however, models may be developed or learned in such a way that their goal representations do not allow reasonable multitasking (i.e., a kind of "impolite" goal representation)—as a simple example, models that only execute a single goal for minutes or hours on end. To this end, this work proposes two heuristics that attempt to further constrain goal representations for facilitation of multitasking and task interleaving:

- The iterating heuristic. Models of complex tasks typically run not for a short burst of time but rather for an extended period of time, implying that at some point these models iterate or repeat sections of their procedures. (For instance, 100 sec of task execution at 50 msec per rule would require 2,000 rule firings; even complex models rarely have near this many rules, but rather iterate on a core set of rules.) The iterating heuristic posits that at any point in which the model iterates on part of its rule set (i.e., by returning to a previously fired rule), the model should allow for task switching at this point by creating a new goal to perform the next iteration of the procedure (as opposed to resetting the state of the current goal).
- The blocking heuristic. Most models of complex tasks involve some amount (if not a great deal) of perceptual–motor activity, and in some cases the production system is forced to wait for a perceptual–motor action to complete before proceeding. The blocking heuristic states that when a model expects to wait a "significant time" for the completion of an action—for example, a hand movement from keyboard to mouse—the model should create a new goal for the next action, thus allowing another task to intercede during the waiting period. At this time, our work unfortunately cannot provide a clear definition of "significant time"; it is very possible that time periods deemed "significant" depend closely on the task at hand and the urgency of all active tasks. Nevertheless, this heuristic has been used in previous work on driver distraction—for example, allowing steering to intercede during a hand movement from the steering wheel to a secondary task device (Salvucci, 2001b)—and, even as an underspecified heuristic, can provide guidance for the granularity of goal representations while modeling.

To emphasize, these heuristics are not intended to be comprehensive or exhaustive, nor do they fully specify a goal representation for a given task. Instead, the heuristics are intended to constrain the space of possible goal representations for tasks such that individual task models integrate well with the general executive. In other words, the proposed approach to multitasking can be viewed as an integration of both an executive control module and a set of general guidelines for modeling individual tasks. The studies of driver multitasking in the next section will illustrate these ideas and also contain several examples of the use of these heuristics to guide modeling of individual tasks.

2.2.5. Summary of the general executive
In essence, the general executive can be characterized by four core ideas:

1. Multiple goals can be "in play" at one time, all competing for execution time on the cognitive processor (unlike current ACT–R theory).
2. However, only one goal can be executed by the cognitive processor at any given time (like current ACT–R theory).
3. Two heuristics guide when to switch away from this goal: (a) switch after completing an iteration of the procedures that achieve the goal, and (b) switch after requesting a perceptual or motor action that requires substantial waiting time.

4. When switching away from this goal, the general executive switches to and executes the most urgent goal (i.e., the goal most due or overdue). Goals can specify when subsequent goals become due on creation of these new goals.

Although this article centers on an instantiation of the general executive in the ACT–R cognitive architecture, the core ideas should generalize well, in whole or at least in part, to other architectures and cognitive theories.

3. Studies in driver multitasking

As described earlier, our previous successful efforts in modeling driver behavior in the ACT–R architecture all relied on customized executives for driver multitasking. Given the proposed general executive, this section describes three studies in which the mechanism is applied to integrating the driving subtasks of control and monitoring (Study 1), integrating driving with the secondary task of tuning a radio (Study 2), and integrating driving with the secondary task of dialing a phone number (Study 3). These studies have two overarching goals. First, the studies aim to replicate some of the findings in our previous work to ensure that the general executive can account for some of the same aspects of driver behavior as did the customized executives in previous models. Second, the studies attempt to extend previous work by accounting for new results related more directly to driver multitasking—specifically, results that elucidate when and how drivers switch tasks.

3.1. Study 1: Control and monitoring

The task of driving actually comprises a number of component subtasks, all critical to safe driving. Perhaps the most obvious and recognizable component subtask is that of *control*: lateral control (i.e., steering) to maintain a central position in the lane, and longitudinal control (i.e., acceleration and braking) to maintain a safe speed, distance, or time headway from nearby hazards. Control clearly requires attention and encoding of the visual environment, with the focus of attention centered on the lead vehicle or upcoming segments of the lane or both. However, safe driving requires additional situation awareness of surrounding vehicles (or other obstacles), bringing up a second critical subtask of *monitoring*. Although monitoring is not as critical as control in an immediate sense, it provides the driver with knowledge of her or his surroundings, thus enabling decision making or emergency maneuvers when necessary (e.g., normal or emergency lane changes); thus, the more time that can be devoted to monitoring, the more accurate the driver's mental model of the immediate surroundings. This first study explores the integration of control and monitoring in a multilane highway environment, such as that shown in Fig. 3, allocating as much attention as possible to monitoring without detracting from the immediate task of safe control.

3.1.1. Empirical study

The human driver data with which to validate the model come from the empirical validation of the original driver model (Salvucci, Boer, & Liu, 2001). This study was conducted in a

Fig. 3. Sample screen shot of the multi-lane highway driving environment.

fixed-base driving simulator with a simulated multilane highway environment and moderate (automated) traffic. The human data comprise a total of 11 participants driving in this environment and generating a total of 311 km of driving data. The resulting data protocols contain a detailed snapshot of the environment sampled at roughly 13 samples/second; each sample includes information about the driver's controls (steering angle, pedal depression, etc.), the driver's vehicle position (lateral position in lane, etc.), and the position of all other vehicles in the environment. Critically, the study included the collection of eye-tracking data that sampled the driver's point of gaze during navigation; these gaze data are crucial to elucidating the interaction of control and monitoring and the switching between them.

3.1.2. Model development

The component models of control and monitoring come from the most recent driver model (Salvucci, in press; an update of the original model in Salvucci et al., 2001). The control model is based on a straightforward perception–action control law (Salvucci & Gray, 2004) that updates steering based on the perceived visual angle to two points: a *near point* directly in front of the vehicle that guides position in the lane, and a *far point* in the distance (e.g., the vanishing point of a straight road or the tangent point of a curved road) that guides steering into the upcoming roadway. This control law works well for both straight and curved roads and also generalizes easily to lane changes. A similar control law updates the depression of the accelerator and brake pedals based on time headway to the lead car. These control laws are incorporated into a set of ACT–R production rules, shown in Table 2, that visually encode the near and far points, compute the necessary values, and generate the motor actions for the steering wheel and pedals. Although many other control models use continuous mathematical functions or control-theoretic approaches (e.g., Hess & Modjtahedzadeh, 1990; Hildreth, Beusmans, Boer, & Royden, 2000), the ACT–R control model, due to its implementation as a production system, necessitates discrete updates of control. Thus, the model updates control at a periodic rate dictated by the number of productions fired (given ACT–R's 50-msec firing rate), and the re-

Table 2
Control and monitoring goals and production rules

Control
    Attend-near: locate near point (if necessary)
    Process-near-attend-far: note near point information, locate far point
    Process-far/car: steer and accelerate for road point or lead car as appropriate
    Done-unstable: if unstable, locate near point and set immediate goal of Control
    Done-stable: if stable, set new goal of Control with delay $D_{control}$
Monitor
    Monitor-lane*: choose random lane and locate if vehicle present
    Done-process-car: note vehicle information, set new goal of monitor
    Done-no-car: set new goal of monitor

Note. Special-case rules that do not affect multitasking have been omitted.

sulting rule set generates updates at a rate of one update per 150 to 200 msec (200 msec for the first iteration after switching to control, 150 msec for subsequent iterations until switching away).

The monitoring model randomly samples the visual environment, which in this case is a two-lane highway with moderate traffic. On each iteration, the model randomly chooses, with equal probability, a lane (left or right) and direction (forward or backward) in which to glance for other vehicles; forward glances are simply through the main view, and backward glances are directed to the rearview mirror. If a vehicle is found, the model notes its lane and direction as well as distance from the driver's vehicle; this information can then be used when deciding whether to change lanes and can, for instance, help to recall a vehicle in the "blind spot" even if the vehicle cannot presently be seen. One iteration of this monitor goal requires 100 msec for the firing of two production rules, one that selects and finds a visual object for a chosen lane and direction, and one that notes the presence or absence of a vehicle in a declarative chunk. The monitoring model's goals and production rules are included in Table 2.

The integration of the control and monitoring models is a straightforward process with the general executive. To initiate driving, the model includes one production that starts both the control and monitor goals. After this point, each task is scheduled and executed by the general executive with no explicit knowledge of the other task. Not surprisingly, there is some communication between the tasks with respect to the monitoring information in declarative memory: When the driver's vehicle approaches the lead vehicle within a certain time headway, the control goal initiates an attempt at a lane change, and this decision cycle can retrieve memories of past monitor goals to check for other vehicles that may hinder the lane change. Nevertheless, the processes do not communicate with respect to multitasking, leaving the work of switching between tasks completely to the general executive.

For both the control and monitoring models, the iterating heuristic applies to the representation of goals: In both cases, the model iterates by creating a new goal of the same type to repeat the cyclic process. By default, the integrated model then would alternate between control and monitoring, interleaving one iteration of one goal with one of the other as illustrated in Fig. 2(a). However, we might expect that the switching between tasks would be at least somewhat dependent on the situation: In situations of difficult control, the control task would dominate and perhaps keep control for a longer time; in situations of easy control or a stable environment, the control task could allow monitoring to occupy more time, given that the more time taken by monitoring, the better the overall situation awareness (e.g., longer looks to estimate velocity rather than simple position). To this end, the model introduces two ways to quantify these factors. First, it includes a *control stability threshold* that indicates how "stable" the external environment should be before the control goal gives up control. The determination of stability, as defined in the most recent driver model (Salvucci, in press), measures whether the vehicle's lateral position is "close enough" to the lane center and whether its lateral velocity is "stable enough" and not moving too quickly side to side; specifically, the three control values that determine lateral position and velocity (near-point position, near-point velocity, and far-point velocity) have constant-value thresholds that, when below all thresholds, define the vehicle as stable. When the environment is not stable, the model performs the next control iteration immediately by setting the delay time to the special *now* value. When the environment is stable, the model includes a *control delay time* that indicates how long the vehicle can go without control until it would be necessary to return. Thus, the interleaving of control and monitoring in fact resembles Fig. 2(b), with control behaving like *task1* in the figure, except that the control delay time will be estimated to best account for driver behavior.

A total of three parameters were estimated for this integrated model. One parameter is associated with the general executive, namely the s_{gq} noise parameter; this was estimated at .075 and kept constant for this study and the two studies that follow. The other two parameters relate to the control model: the control delay time $D_{control}$, estimated at 500 msec and kept constant across studies; and a control stability threshold F_{stable} that scaled the threshold in the original driver model by a constant factor, estimated at .71. For analysis, model predictions were collected by running three 10-min simulations in the same multilane highway environment. Because the same environment was used for human data collection and model simulation, the model generates behavioral protocols identical to those of human drivers, and thus its behavioral data can be analyzed in exactly the same way. The model, like human drivers, also generates gaze data through ACT–R's vision module (see Byrne & Anderson, 2001) integrated with the EMMA module that translates movements of visual attention to observable eye movements (Salvucci, 2001a).

3.1.3. Results

3.1.3.1. Aggregate measures of task switching. The most informative indicator of task switching between control and monitoring arises in the form of driver gaze, noting where drivers direct their overt visual attention as eye movements to visual objects in the environment. One aggregate measure of task switching, then, examines the proportion of gaze time spent on different visual regions in the environment. Fig. 4 shows this measure for several regions in the environment: the near region of the road, the far region of the road, the lead car, and other cars (forward only) in this lane; these same regions in the other lane; the rearview mirror; and vehicles in the oncoming lanes. Gazes serving the function of control are represented by gazes to the current lane in the near region of the road, far region of the road, and lead car. The figure

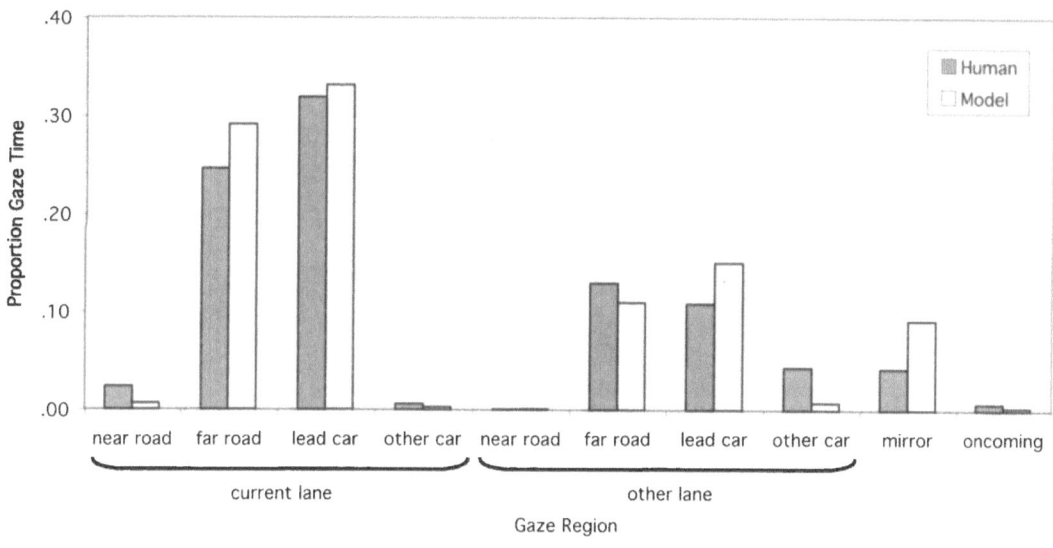

Fig. 4. Study 1: Aggregate proportion gaze times for visual regions in the multi-lane highway environment.

clearly shows that drivers directed the majority of their gazes to these regions, particularly the far road and the lead car. The model, using these points for control as well, exhibited very similar gaze times on these regions. Gazes serving the function of monitoring are represented by gazes to the other lane and the mirror. Drivers spent approximately half the time on these monitoring regions as compared to the control regions. Also, the drivers gazed at the rearview mirror roughly 5% of the time to monitor cars in back of the vehicle. Again, the model predicted very similar gaze times on these regions, albeit with slightly high predictions for the mirror. The default model used here assumed equal probabilities for forward and backward monitoring, so the addition of an extra parameter for weighting forward versus backward gazes could have more closely estimated the mirror gaze time. Nevertheless, the overall model fit to the human data remained quite good, $R^2 = .95$, $RMSE = .03$. Thus, the model nicely accounts for the aggregate effects of task switching on overall gaze time, specifically the amount of time spent on control versus monitoring gazes.

3.1.3.2. Detailed measures of task switching. Although these aggregate measures show some evidence of proper task switching, they do not speak to exactly when drivers switch between tasks; as pointed out by Gray and Schoelles (2003), models that capture aggregate measures of task switching do not necessarily correctly predict switching behavior at a lower level. To this end, we can examine when drivers switched between control and monitoring by measuring the probability of switching over time, where a "switch" is defined by a shift of gaze from the control visual regions to the monitoring visual regions and vice versa. To perform this analysis, all gazes were first classified as control or monitoring gazes, based on the regions necessary for their respective goals: Gazes on the same lane (forward only) were classified as control gazes; gazes on other lanes or in the rearview mirror were classified as monitoring gazes; and all other gazes were classified in a catch-all "other" category. Next, all one-sample gazes interrupting a continuous gaze at a particular region ("blips" in the eye-movement data, often from eye blinks or other data noise) were included in the larger gaze. Finally, the resulting gazes were grouped together in subsequences by task (control, monitoring, or other), and switch probability distributions over time were computed from these subsequences at ¼-sec intervals. This analysis excluded any samples that were part of a lane-change maneuver to remove any ambiguity of the meanings of "current" and "other" lane.

Fig. 5(a) shows the resulting switch probability distributions for monitoring, including the human data (solid line) and model predictions (dashed line). Human drivers were most likely to switch after 250 to 500 msec of monitoring, with a sharp drop-off thereafter and very few switches after 1 sec. It is important to note that the peak in this distribution is not simply a function of the time needed to visually encode another car's position (which can be done peripherally and would require 100 msec for ACT–R); drivers attend to the monitoring task as long as possible, knowing that longer looks can provide more information (e.g., velocity) and more accurate information. The model reproduced this distribution, $R^2 = .95$, $RMSE = .04$, primarily as a function of the control delay time in the control model: Because a stable control goal requests a delay time of 500 msec (as estimated in parameter fitting), the integrated model can run several iterations of the monitoring goal until this delay time expires and the model returns to control. Thus, the model may monitor several vehicles during this delay, or may even monitor the same vehicle for an extended period of time. The control delay time also results in very few switches after 1 sec, because the urgency to switch back to control increases as monitoring runs longer past the 500 msec delay time.

Fig. 5(b) shows the analogous distribution for control, and here we see a very different pattern: The highest switch probability for human drivers fell in the 0 to 250 msec range, and the probabilities dropped steadily from this point in a smooth manner. It was not uncommon for drivers to perform control for 1 to 2 sec before switching, indicating points of difficult driving during which the driver maintained interrupted control. The model nicely reproduced the overall trends in the data, $R^2 = .97$, $RMSE = .02$. In the model, the control goal relinquishes control only when the vehicle achieves the stability threshold, thus generating a number of long control times in the 1 to 2 sec range. However, usually the model required only a short period of control during stable driving, and thus the model, like human drivers, exhibited the highest switch probability in the shortest duration range.

3.2. Study 2: Radio tuning and driving

Although driving itself can be decomposed into subtasks, another common aspect of driver multitasking involves the integration of driving with some secondary task, such as the use of an in-vehicle device like a cellular phone or navigation device. The burgeoning use of such in-vehicle devices has begun to pose a serious safety risk: A recent study determined that driver distraction and inattention is now the leading cause of crashes, ahead of even speeding and alcohol (Hendricks, Freedman, Zador, & Fell, 2001). This second study explores the driver multitasking that occurs when integrating the primary driving task with the secondary task of tuning a radio device. The study is based on empirical data collected by Sodhi, Reimer, and Llamazares (2002) that examined driver eye movements while performing various secondary tasks.

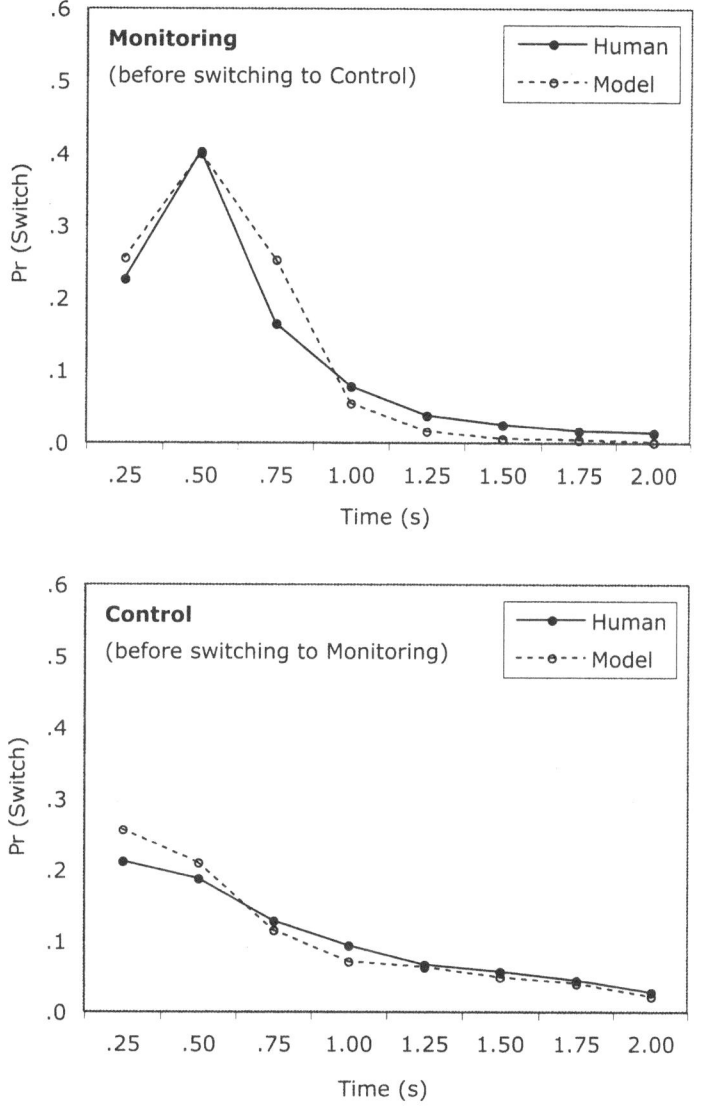

Fig. 5. Study 1: Probability of switching after a given time for (a) monitoring and (b) control.

3.2.1. Empirical study

In the Sodhi et al. (2002) study, participants drove an instrumented vehicle on a semirural road at a self-selected natural speed. Although driving, participants were occasionally asked to perform a variety of secondary tasks, including the one of interest to this study, namely, the radio-tuning task in which the participant turned on the radio and tuned the station to 1610 AM. Throughout the experimental drive, participants' eye movements were monitored using a head-mounted eye tracker sampling gaze direction at 50 Hz. These eye movements were subsequently processed and classified as glances either to the roadway (for the driving task) or the radio (for the secondary task). Sodhi et al. (2002) provided aggregate analyses as well as individual protocols illustrating drivers' behavior during the secondary tasks; this article focuses on the results that best illustrate the task-switching aspects of driver behavior, namely, time spent on one task before switching to the other as elucidated by the eye-movement protocols.

3.2.2. Model development

The component models for this study include the primary task of control and the secondary task of tuning a radio to a desired station. The control model comes, not surprisingly, directly from Study 1, with production rules and declarative representations simply ported over from the previous study into this study. However, because the driving environment differed in this study, the model was placed on a single-lane roadway with a single lead vehicle to help the model drive at a constant speed. It is important to note that the monitoring portion of the driver model became irrelevant in this environment—with no vehicles in other lanes and no possibility of lane changing—and thus the model only required the control component of the driving model used in the first study.

The radio-tuning model, derived from a task analysis of tuning a standard car radio, is outlined in Table 3, which lists the goals that make up the model as well as the individual production rules associated with each goal. Unfortunately, the original study did not specify the initial state of the radio, so the model assumes that drivers select a station in two steps: (a) holding down the station-advance button for 2 sec to jump approximately to the station, and (b) press-

Table 3
Radio-tuning goals and production rules

Tune-Move
    Find-radio: locate radio (peripherally)
    Encode-radio: fixate and encode radio
    Move-to-radio: move right hand to radio, set new goal of Tune-Power-On
Tune-Power-On
    Find-power-button: locate power button
    Encode-power-button: fixate and encode power button
    Press-power-button: press power button, set new goal of Tune-Start-Jump
Tune-Start-Jump
    Find-advance-button: locate advance button
    Encode-advance-button: fixate and encode advance button
    Hold-advance-button: press advance button, set new goal of Tune-Monitor-Jump
Tune-Monitor-Jump
    Find-display: locate station display
    Encode-display: fixate and encode display
    Hold-advance-button: before 2 sec, set new goal of Tune-Monitor-Jump
    Release-advance-button: after 2 sec, release advance button and set new goal of Tune-Adjust
Tune-Adjust
    Find-display: locate station display
    Encode-display: fixate and encode display
    Press-advance-button: press advance button, set new goal of Tune-Return
Tune-Return
    Home-hand: move right hand to steering wheel, terminate goal

---

ing the advance button one last time to adjust for error in this jump and select the exact station. (These assumptions were made a priori before any simulations were performed and were not modified thereafter.) Referring to the table, the Tune-Move goal visually locates the radio and moves the right hand to its general location, then Tune-Power-On locates the power button and presses this button to activate the radio. Next, Tune-Start-Jump presses the station-advance button and Tune-Monitor-Jump iterates to watch the display and complete the jump after 2 sec. Finally, Tune-Adjust presses the advance button one last time, and Tune-Return moves the right hand back to the steering wheel.[2] The model incorporates the iterating heuristic in that the Tune-Monitor-Jump goal iteratively creates the same goal until the appropriate time. The model also incorporates the blocking heuristic in that each hand movement or button press is considered a significant event (requiring visual fixation and then movement), which must wait for the physical action to complete and thus allows another goal (i.e., control) to intercede.

As in Study 1, the creation of the integrated model for control and radio tuning becomes a straightforward task, given the general executive. On starting up, the model initially runs only a control goal to navigate the construction-zone environment. To initiate tuning, the model simply adds the tuning goal to the goal queue, thus allowing tuning to interleave with driving as dictated by the general executive. The general executive here uses the same noise value estimated in Study 1, and the control model uses the same control delay time (500 msec) as in Study 1. However, because the studies used different driving environments, we expect that human drivers exhibited different abilities or tolerances in the environments. To account for this, the model incorporated constant scaling factors for three sets of parameters, namely for the amount of steering change, the amount of pedal-depression change, and the control stability threshold; these scaling factors were estimated to produce the best fit to the empirical data for both Studies 2 and 3, with final values of 0.7, 0.4, and 2.5, respectively. The integrated model was run in three driving simulations with eight tuning trials per simulation, spaced 20 sec apart.

3.2.3. Results

In Study 1, the measure most illustrative of drivers' task-switching behavior was the probability of switching between tasks as indicated by eye-movement data. Fortunately, Sodhi et al. (2002) reported these same switch probabilities for drivers performing the radio-tuning task, and thus we can validate the integrated model's predictions through comparison with these data. Fig. 6(a) shows the switch probability distributions for radio tuning, including the human data (solid line) and model predictions (dashed line). The human drivers switched most often in the 0.6- to 1.0-sec range, with fewer switches in the first 0.0 to 0.4 sec, even fewer switches after 1.0 sec, and almost no switches after 1.6 sec. As before for monitoring, these distributions indicate a balance between keeping a gaze on the radio (primarily to confirm the changing display) and ensuring that control is able to intercede at regular intervals. The model nicely reproduced this switch probability distribution, $R^2 = .92$, $RMSE = .03$; the control delay time (kept constant between studies) allowed the model to concentrate on tuning initially, thus leading to a few quick switches before 0.4 sec, but also forcing the model to interrupt and switch to control as time passed. Switches around 0.6 to 1.0 sec were primarily due to the initial Tune-Move and Tune-Power-On goals running in sequence, without a switch in between, for a total duration of approximately 0.8 sec; switches around 1.4 sec were primarily due to the Tune-Adjust

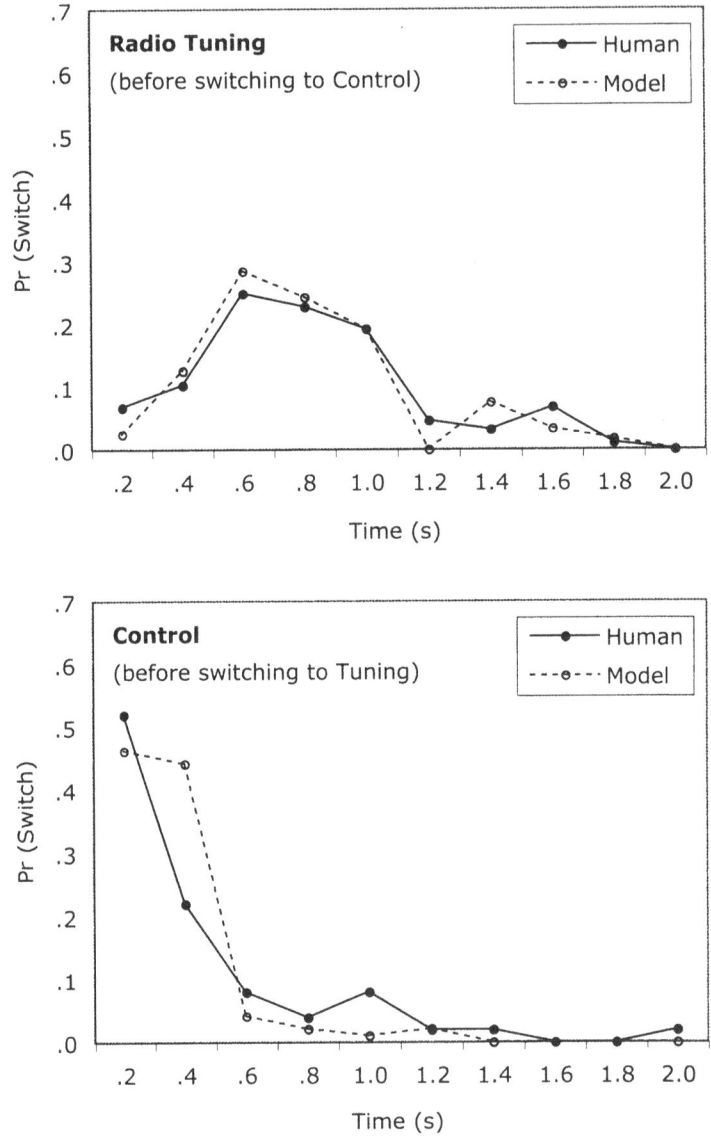

Fig. 6. Study 2: Probability of switching after a given time for (a) radio tuning and (b) control.

and Tune-Return goals running in sequence and waiting for the hand to return to the steering wheel before terminating the goal.

Fig. 6(b) shows the analogous distribution for control, with a pattern very similar to that in Study 1—namely, that the switch probability was rather high for short times and tailed off for higher times. The model again fit the data well, $R^2 = .81$, $RMSE = .08$, albeit with a discrepancy in the 0.2- to 0.4-sec range. Again like Study 1, the model under normal conditions often produced stable control with single control updates, leading to frequent short switch times; under less frequent, difficult driving conditions, or when the vehicle became unstable, the model focused on control and exhibited longer switch times.

3.3. Study 3: Phone dialing and driving

Study 3 also explores driver multitasking in the context of driver distraction. However, whereas Study 2 examined a highly visual radio-tuning task, Study 3 examines a primarily nonvisual task of dialing a familiar 10-digit phone number on a numeric keypad with no visual feedback. Several of our previous efforts have shown how the ACT–R driver model can predict various effects of phone dialing on driver performance (Salvucci, 2001b, 2002; Salvucci & Macuga, 2002). This study presents new empirical results that delve deeper into exactly when drivers interleave driving and dialing, and it also demonstrates how the general executive can reproduce these results by integrating separate models of the two tasks. The study uses the same highway environment as Study 2, namely, a realistic one-lane highway with a lead vehicle, with one exception: The lead vehicle accelerates and decelerates in an abrupt random manner similar to what might occur in a construction zone (see Salvucci & Macuga, 2002). As a result, we can analyze both the aggregate effects of multitasking on driver performance, and also the detailed effects of multitasking that elucidate when drivers switch between driving and dialing.

3.3.1. Empirical study

The human driver data for the driving and dialing task were collected from drivers navigating the construction-zone environment in a small-scale driving simulator, specifically a desktop system with a force-feedback Logitech Wingman® steering wheel and integrated pedals. In the experiment, each participant was first given general information about the driving environment and allowed 5 min of practice driving. Then, the participants wrote down four of their most familiar 10-digit phone numbers along with a name or phrase associated with each number. The experiment continued with three phases: pretest, main, and posttest. In the main phase of the experiment, participants drove in the environment and were occasionally asked to dial a phone number: The experimenter asked the participant to dial and simultaneously hit a special start key on the keyboard; the participant then dialed the number on the numeric keypad of the keyboard, and finally terminated the number by pressing *Enter* on the keypad. In the pretest and posttest, the same protocol was followed, but the simulation was turned off, and the participant had only to dial the number (without driving). Each participant performed a total of 32 dialing trials (8 for each of the four phone numbers) in the main driving test and 32 trials combined in the pretest and posttest (with 16 trials each). In all, data from 10 participants were collected for the study. Of these, 3 participants failed to dial 90% of the numbers correctly and were omitted from further analysis. The data from the remaining 7 participants include 51 km of driving data.

3.3.2. Model development

The component models for this study include the primary task of control and the secondary task of dialing a phone number. The control model was taken from Study 2, and because both studies use essentially the same driving environment, the model could be directly imported for this study with no changes to the model, including parameter settings.

Table 4
Phone-dialing goals and production rules

Dial-Move
     Find-keypad: locate numeric keypad
     Look-at-keypad: fixate and encode keypad
     Move-to-keypad: move right hand to keypad, set new goal of Dial-Prepare
Dial-Prepare
     Recall-first-block: retrieve first block
     Start-dial: set new goal of Dial-Block
Dial-Block
     Recall-first-number: retrieve first number of block
     Type-number-recall-next: type current number, recall next number in block
     Recall-next-block: retrieve next block
     Do-next-block: if there is a next block, set new goal of Dial-Block
     Done-blocks: if there is no next block, continue
     Press-enter: type *Enter* termination key, set new goal of Dial-Return
Dial-Return
     Home-hand: move right hand to steering wheel, terminate goal

---

The dialing model derived from a task analysis of dialing a 10-digit phone number, resulting in the goals and production rules outlined in Table 4. First, the Dial-Move goal locates the numeric keypad and moves the hand from the steering wheel to the keypad, assumed here to be equivalent to a move from the home row of a standard keyboard to the numeric keypad as implemented in ACT–R. Next, the Dial-Prepare goal prepares the dialing procedure by recalling the declarative chunk representing the first block of numbers; the model segments the phone number into blocks of 3, 3, and 4 digits corresponding to the typical segmentation *xxx-xxx-xxxx*. Then, the Dial-Block goal iteratively dials each block by recalling and typing each digit and recalling the next block. When the retrieval of the next block fails and thus dialing is complete, the model presses the *Enter* key and moves the hand back to the steering wheel. Finally, the Dial-Return goal moves the right hand back to the steering wheel and terminates the dialing goal. The iterating heuristic applies for the Dial-Block goal, which repeats for each of the three blocks in the phone number. The model assumes that the keystrokes that compose the actual dialing in the Dial-Block goal are not "significant" pauses, and thus the blocking heuristic does not apply. However, the model does consider hand movements between steering wheel and keypad significant pauses, and thus the blocking heuristic applies for the Dial-Move and Dial-Return goals.

    Typing times are particularly critical for the dialing model's performance, but also susceptible to large variability in the human data, and thus it was important to ensure that the model's typing matched reasonably well with that of the human participants. To this end, the execution time for keying a digit on the keypad was set to the average keystroke time found in the empirical study, 260 msec, minus the 50-msec cognitive initiation time needed for a production firing, resulting in a 210-msec motor time. In her TYPIST model, John (1996) found that a keystroke motor time of 230 msec worked well for typists at speeds of 30 gross words per minute (gw/min), but this time can decrease significantly for faster typists—for example, to 170 msec for a 60-gw/min typist. Thus, the average keystroke motor time found here falls nicely within her reported range of typical times. In addition, the model uses John's (1996) assumption that the cognitive processor waits for the completion of each keystroke before firing a new production rule.

As in Studies 1 and 2, the creation of the integrated model for driving and dialing becomes a straightforward task, given the general executive. As in Study 2, the model initially runs only the control goal to navigate the environment, then adds the dialing goal to initiate dialing. The general executive again uses the same noise value as the other two studies. The control model again uses the same control delay time, and because the Study 3 driving environment is identical to that in Study 2, this study imports all other control parameters from Study 2. The integrated model was run in five driving simulations with eight dialing trials per simulation spaced 20 sec apart, totaling 20 km of driving data. To avoid the possibility of the driver model overtaking the lead car (e.g., if it braked too late to avoid a crash), the model's vehicle was constrained to a minimum following distance of 10 m behind the lead vehicle. The model was also run in five simulations without driving to generate baseline performance on the dialing task.

3.3.3. Results

3.3.3.1. Aggregate effects of task switching. The human and model data arising from this study include a host of information elucidating the multitasking behavior with driving and dialing. Before moving to a detail analysis of task switching, it is important to ensure that the general executive produces the same effects observed in previous studies of driver distraction—namely, aggregate effects of dialing on driving, and also effects of driving on dialing. First we consider aggregate effects of driving on dialing. Fig. 7(a) shows the mean dialing times in the baseline condition, representing the data collected during the pretest and posttest with no driving task, and in the driving condition, representing the data collected for dialing while driving. For the human drivers, driving had a significant effect on dialing time, $t(6) = 3.40$, $p < .05$; perhaps surprisingly, this difference is not large, with driving adding only 0.55 sec to the mean dialing time (4.46 sec baseline, 5.01 sec driving). The model produced a very close fit to these data, including the observed increase due to driving. In essence, driving does take some time away from execution of the dialing sequence, but at the same time, the control updates occur quickly and can be smoothly interleaved with dialing such that the total time added remains relatively small. The model exhibited almost no variability in the baseline condition but a small amount in the driving condition, where vehicle stability played a role in allowing or disallowing the model to switch away from control.

Next we consider aggregate effects of dialing on driving. Fig. 7(b) shows two measures of driver performance: lateral deviation, computed as a root-mean-squared error between the vehicle's current lateral position and the center position of the lane (see, e.g., Salvucci, 2001); and speed deviation, computed as a root mean squared error between the vehicle's current speed and the speed of the lead vehicle. As in such analyses in previous studies, these measures were computed both during normal driving and while dialing, where the latter included a 5-sec window after completion of dialing to account for subsequent corrections in control. For lateral deviation, human drivers exhibited a significant effect of the dialing task, $t(6) = 2.99$, $p < .05$. The model also exhibited this effect; the effect size was larger

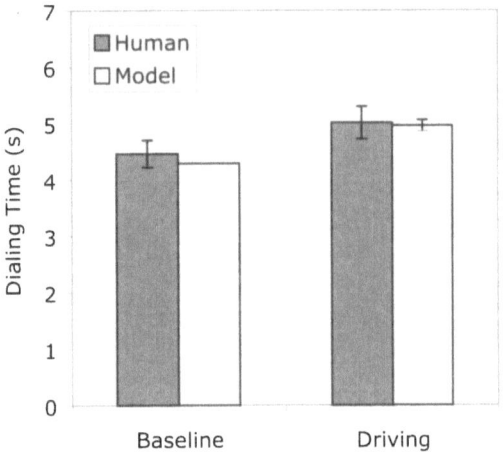

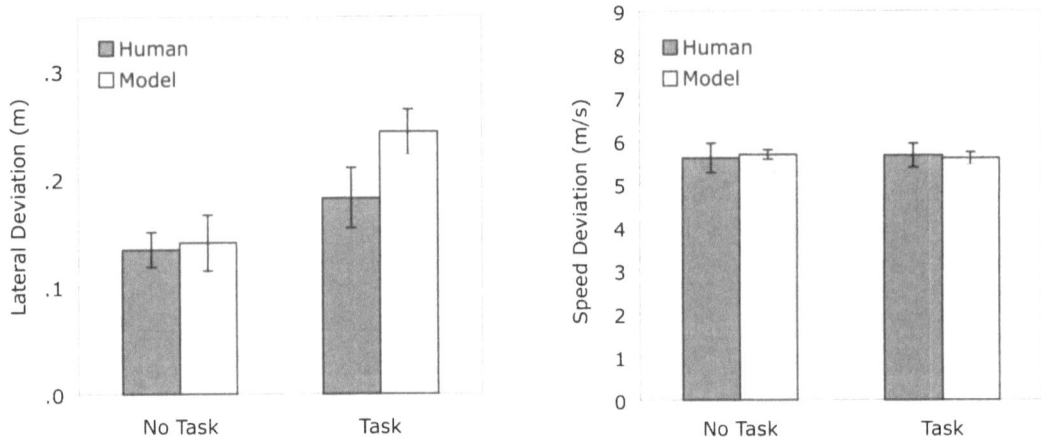

Fig. 7. Study 3: Aggregate effects of (a) driving on dialing as measured by dialing time, and (b) dialing on driving as measured by lateral deviation and speed deviation. Errors bars represent standard errors.

for the model than for human drivers, although there was also considerable variability in both model and human data. Interestingly, there was no significant effect of driving on the speed deviation measure for the human drivers, $t(6) = .35$, $p > .5$; the car-following aspect of the driving task turned out to be quite difficult for the drivers, leading to a fairly high mean speed deviation (over 5 m/sec or roughly 11 mph), and thus the presence of the secondary task did not add significantly to this difficulty. The model, experiencing the same difficult car-following as the human drivers, also showed no effect of the secondary task. This high-lights an important finding for the model and for the general executive: Multitasking may show distraction effects for some measures and not for others, and the general executive nicely predicts the significant effects and, just as importantly, the lack of significant effects for appropriate measures.

3.3.3.2. Detailed measures of task switching. Although this overall picture of dialing provides some clues as to the efficient interleaving of driving and dialing, perhaps the most important validation for the general executive arises in analysis of exactly when drivers switched between dialing and driving. In the radio-tuning task, we could analyze driver eye movements to find when drivers switched, because the task necessitated visual fixation on the radio for successful execution. In the phone-dialing task, drivers had no need to take their eyes off the road because the interface (i.e., the numeric keypad) gave no visual feedback. Thus, the phone-dialing task required a different approach to analyzing task-switching behavior. For this purpose, our study uses *key delays* to elucidate driver task switching, namely, the time between one key press and the next during the course of dialing a phone number. The baseline condition (dialing alone without driving) provides baseline key delays without interference from driving; the driving condition (dialing while driving) provides delays that include interference from driving. By comparing these baseline and driving conditions, we can determine where extra time was needed in the driving condition and thus where, presumably, drivers interleaved control with dialing.

For the human drivers, Fig. 8(a) shows the mean key delays computed as the time preceding each key press, including the 10 key presses for the phone number and the final *Enter* key press. The first key press showed a very significant effect of driving, $t(6) = 4.17$, $p < .01$, and the delays for both baseline and driving were much longer than the other delays due to initiation of the dialing (e.g., hand movement to the keypad) and because the driving condition included extra driving time to ensure stability of the vehicle before starting to dial. Also, the key presses that initiate new blocks within the phone number (Keys 4 and 7, for a phone number of the form *xxx-xxx-xxxx*) showed a significant effect of driving, $t(13) = 3.12$, $p < .01$; again, this difference indicates the extra time needed to control the vehicle to an acceptable stability. However, both the intermediate key presses (Keys 2–3, 5–6, 8–10) and the final *Enter* key press (Key 11) showed no significant effects of driving, $p > .30$. Thus, the human drivers seemed to interleave iterations of control at the block boundaries, but not within blocks, except for the final key press, which required more time than the intermediate key presses but still showed no significant difference while driving.

Fig. 8(b) shows the analogous graph of key delays for the model. Overall the model reproduced many aspects of the human driver data, $R^2 = .96$, $RMSE = .10$, including: (a) the longest baseline time for Key 1 and longer baseline times for Keys 4, 7, and 11; (b) effects of driving for the first key press and for the block-starting key presses, namely, Keys 1, 4, and 7; (c) no effects of driving for the intermediate key presses and the final key press. Like the human drivers, the model exhibited effects on the block boundaries because of the extra time needed to control the vehicle; these block boundaries coincide with the goal boundaries of dialing a block of numbers, and thus this result falls directly from the goal structure in the model. Although the effect of driving for Keys 4 and 7 was slightly larger than those for the human drivers, there was a fair amount of variability in the driving condition because of the "randomness" of how these boundaries coincide with different driving situations (e.g., whether the boundary occurred during a stable straight road segment or a difficult curved

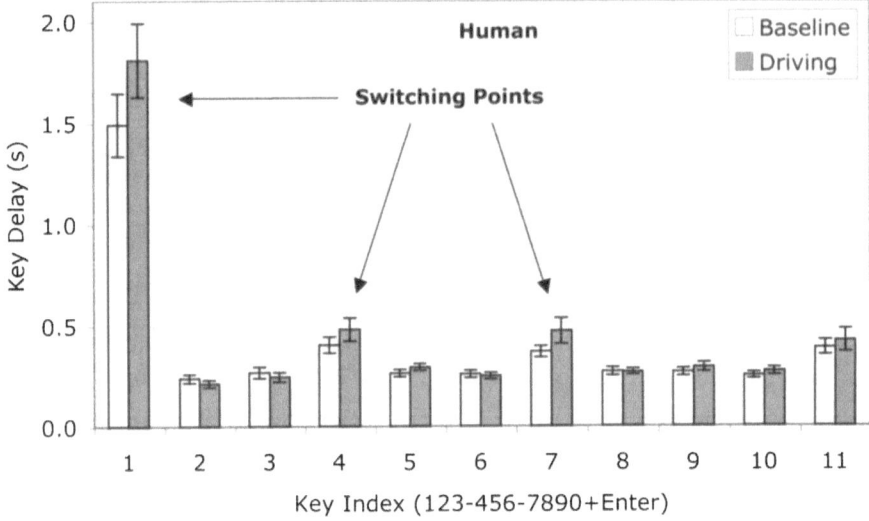

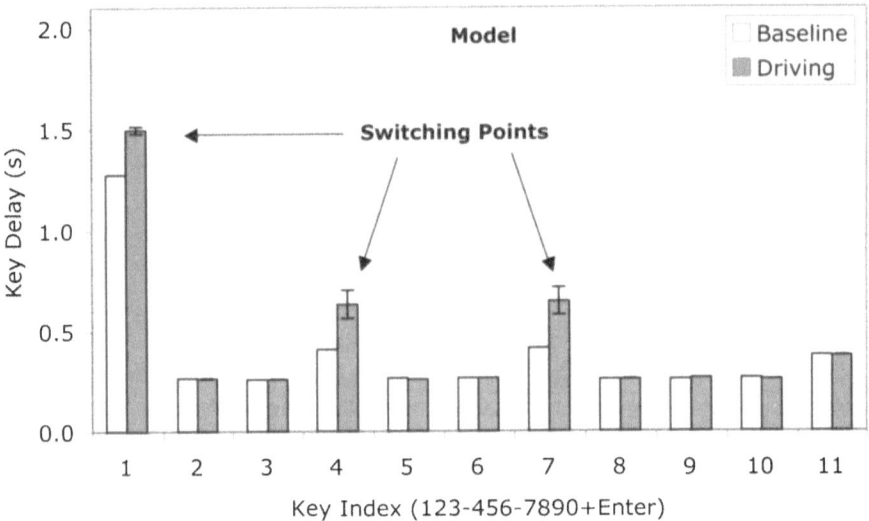

Fig. 8. Study 3: Task-switching points as illustrated by key delay times for (a) human drivers, and (b) model simulations. Errors bars represent standard errors.

segment). The model also somewhat underpredicted the delay times for the first key press; one could argue that the model has more "readiness" to start dialing immediately, whereas drivers seemed to exhibit some extra time to initiate the dialing goal. Nevertheless, the model nicely reproduces the trend in the key press data of where drivers switch back from dialing to driving, specifically at the block boundaries.

3.4. Summary and lessons learned

These three studies explored how the general executive can account for various aspects of driver multitasking. In our previous modeling efforts in this domain (e.g., Salvucci, 2001b, 2002; Salvucci & Macuga, 2002), the models used customized executives to exhibit multitasking behavior: The component models managed processes by explicitly passing execution from one to another through the setting of new goals (e.g., driving passing control to dialing, then dialing back to driving). The studies here demonstrate how the general executive provides a more rigorous and plausible description of driver multitasking in three ways as related to the general principles outlined earlier.

First, the general executive as an architectural mechanism provides a more plausible and consistent account of multitasking that more easily generalizes across studies, as compared to previous domain-specific customized executives. In the previous models, the component models could not be easily separated from each other—for instance, because the control model specifically mentioned and passed execution to the dialing model, removal of the dialing model would "break" the control model. Such a strongly unified model might be reasonable after long periods of learning (arguably the control and monitoring models presented here may achieve that level of integration), but component models for individually well-learned skills, such as driving and dialing, can clearly be decomposed and run independently without reliance on the other. Second, to achieve integration with another task model, the component models would require modification to explicitly manage execution with the new model—for instance, for driving and radio tuning, the driving model would require changes to set the goal to radio tuning, and the tuning model would need analogous changes to set the goal back to driving. This would imply that with each new integrated set of tasks, a person would need to learn new rules to manage that specific set of tasks. Again, such rules might sometimes be achievable through learning, but learning specific rules to manage all possible (or reasonable) sets of tasks would lead to an exponential explosion in rules and seems implausible as models account for a fuller scope of human cognition.

Because the general executive handles task management and scheduling, the four models used in the studies (control, monitoring, radio tuning, and phone dialing) do not refer or pass execution to one another, but instead only take care of their own processing. For this reason, the models could plausibly be learned independently of one another and can also run independently if desired. Second, the general executive mechanism remained the same across all studies, and its one parameter (temporal noise) remained constant across all studies. This further confirms our goal of having a single mechanism operate over any set of possible desired tasks. Third, the primary task of vehicle control, common to all three studies, was handled by the same control model across studies. The only caveat to this reuse arose in changes to control-specific parameters due to differences in driving environments in the empirical studies; nevertheless, even these parameters remained unchanged when using the same driving environment (namely, Studies 2 and 3 used the same driving environment and thus used the exact same control model). Most important, the model parameter most critical for multitasking, the control delay time, remained constant across all studies. Thus, unlike our previous studies with customized executives, these studies demonstrate how reuse of both the general executive and component models can greatly facilitate modeling of multitasking and lead to a more plausible account of the integration of independently learned component models.

The second way in which the general executive extends our earlier models of driving relates to the incorporation of time in the multitasking model. The previous models did not reason about time in their multitasking schemes, but rather used one of two methods for multitasking: (a) They defined an estimated probability of switching from this task to another—for instance, the probability of monitoring after one cycle of control updates; or (b) they allowed only a single iteration before switching tasks—for instance, performing a single monitoring goal and then switching back to driving. These two methods could produce only a very limited set of switch probability distributions, namely, an exponential distribution that fades over time (a) or a trivial distribution where all switches occur at the same point in time (b). These two distributions clearly do not match the switch probabilities observed in the driving studies, particularly those in Figs. 5(a) and 6(a). The proposed general executive helps to account for the temporal trends in these data by incorporating time into its task scheduling and allowing tasks to proceed until their time "expires" and they relinquish execution.

The third way in which the general executive benefits previous driving work involves the use of representational heuristics to guide model goal representations. The previous studies, in fact, used the iterating heuristic implicitly by switching only after completed iterative cycles (e.g., after a control update), and they also hinted at the blocking heuristic as part of the guidance for the development of the first dialing model for driving (Salvucci, 2001b). The two heuristics as defined here help to formalize these previous assumptions into two basic heuristics that can guide future model development. In addition, Study 3 in particular demonstrates that goal representations can play a key role in where people switch between tasks. As mentioned earlier, although these heuristics (particularly the blocking heuristic) do not fully specify the desired goal representations, they at least help constrain the space of representations in such a way as to make models more amenable to integrated multitasking.

4. General discussion

The proposed general executive consists of both a computational mechanism and representational heuristics for allowing interleaved execution of multiple task models. The application to the driving domain illustrates how the general executive facilitates modeling of compound continuous tasks and helps to account for both aggregate effects of multitasking and detailed measures of task-switching behavior. Of course, no one domain or one modeling effort can suffice in validating the general executive; further studies in both simple laboratory domains and complex dynamics tasks are needed to flesh out its strengths and weaknesses. However, driving serves as an excellent case study in launching such an effort for the ACT–R architecture: It serves as a representative example of a wide variety of domains that demand dynamic integration of cognitive, perceptual, and motor processes in the context of a complex real-world task. The driving studies here demonstrate that the general executive is at least general enough to account for a variety of aspects of driver multitasking, showing promise and opening up the theory to further validation in other domains.

The development of models for complex tasks often relies on prior development of components of these tasks, and thus model reuse is critical in facilitating both theoretical consistency and practical development. Reuse can occur in a number of ways. The cognitive architecture itself provides the first layer of reuse in that its built-in theories and mechanisms cut across all models with a baseline theory and simulation engine. At times, reuse can provide common representations of information for general tasks—for instance, representations (and perhaps corresponding models) proposed for list memory (Anderson, Bothell, Lebiere, & Matessa, 1998) and analogy (Salvucci & Anderson, 2001) offer a common way of using these general mechanisms *within* larger models. At a higher level, reuse can involve embedding entire models within larger models, including declarative representations and chunks as well as their associated production systems. The studies of driving presented here serve as one example: A model of radio tuning or phone dialing can be developed and run completely by itself, but can also be easily integrated with the driver model to immediately generate emergent predictions of the interactions that arise between the tasks. This type of reuse with the proposed executive should generalize in a straightforward way to a large range of possible models; for example, we might integrate the driver model with Pirolli's (this issue) model of information retrieval to predict interactions of driving with e-mail or Web access (e.g., through currently available "smart phone" devices); we might integrate the driver model with Lewis's (this issue) model of parsing to predict interactions of driving with display reading or conversation; or we might integrate Lewis's model with Anderson's (this issue) model of algebraic symbol manipulation to predict how students might encode instructions or "over-the-shoulder" help while solving algebra problems. Model reuse offers enormous predictive power to cognitive architectures: As individual researchers explore particular issues of driving, language, memory, and so forth, the entire modeling community immediately benefits from the work and can integrate the resulting models, representations, and so forth, into their own efforts.

Besides the critical issue of model reuse, another issue brought to light by the general executive is the need for awareness and reasoning about time. Time clearly serves as an essential aspect of many dynamic complex tasks, and thus it is not surprising that models of these tasks must exhibit awareness of time as part of their behavior. The proposed general executive incorporates an implicit awareness of time through its scheduling mechanism: New goals are temporally ordered on the goal queue, and a model can modify the default ordering through specification of goal delay times. Other modeling efforts have recently begun to explore explicit reasoning about the passage of time for tasks such as interval estimation (Taatgen, van Rijn, & Anderson, 2004), including accounts of the types of errors made during estimation of temporal intervals. These efforts nicely complement the proposed executive and, perhaps in the near future, could replace its now-rudimentary formulation of time with a more rigorous account of temporal reasoning.

Although the general executive focuses currently on internally driven task switching, some task switching can be driven externally by the environment—for instance, when a choice stimulus appears on screen during a manual tracking task. The general executive does not directly speak to this issue, but in fact offers a solution for dealing with such task switches: When the model does sense the external stimulus, it creates a new goal with the purpose of responding to the stimulus, which itself is then interleaved with the primary task. In ACT–R, the sensing of the external stimulus can be performed through use of the "buffer stuffing" (e.g., Fick & Byrne, 2003) in which a new stimulus immediately appears in a visual or other perceptual buffer. Subsequently, a production rule can react to the presence of a new stimulus in the buffer and create a new goal to handle and respond to the stimulus. Thus, although the general executive does not provide a method of handling the environmentally driven stimulus directly, such a method already exists in the ACT–R architecture, and the general executive provides a complementary mechanism that allows a model to respond without necessarily drawing full attention away from the primary task.

The general executive as currently specified certainly does not address all the known issues related to multitasking within cognitive architectures. One significant limitation is that the executive currently has no relation to declarative memory and activation, and thus goals cannot be retrieved or forgotten like other declarative chunks. For our chosen domain of driving, such issues are arguably less relevant: When switching tasks at the subsecond level, presumably goals rarely if ever fall below a threshold of retrieval at which they would fail to be recalled (e.g., a driver forgetting the primary driving goal after switching to radio tuning), at least in part because of continual reminding from environmental stimuli. Nevertheless, I expect that as the general executive generalizes to other domains, issues concerning goals as declarative chunks, such as decay and interference (e.g., Altmann & Gray, 2002), will require modifications to this framework.

Another critical aspect involves the modeling of improvement in multitasking over time (Chong, 1998), or put another way, transitions between the stages of multitasking skill acquisition (Kieras et al., 2000). The initial stage of learning would include the learning of declarative instructions that, over time, shift to more procedural execution (e.g., Taatgen & Lee, 2003). In the final stages of learning, human behavior can become highly skilled such that the parallel streams of cognition, perception, and motor movements are highly optimized and interleaved; such behavior has been modeled in cognitive architectures in smaller task paradigms (e.g., Byrne & Anderson, 2001; Meyer & Kieras, 1997) and larger complex tasks (e.g., Vera, Howes, McCurdy, & Lewis, 2004), and notably Taatgen (2005, this issue) explores how such optimizations may arise directly from instruction learning. The general executive proposed here lies somewhere in between: The skills of driving, radio tuning, and phone dialing are well-learned tasks—well past the instruction stage—and thus the executive manages these production-level compiled tasks, but at the same time the executive does not predict the highly optimized interleaving demonstrated in the recent work mentioned previously. Nevertheless, my hope is that the general executive will generalize to incorporate such mechanisms and help bridge the sometimes large gap between initial learning and extremely skilled execution.

Notes

1. In the actual production-rule syntax, the general executive reinterprets standard ACT–R syntax such that "+goal" adds new goals and "–goal" removes these goals. It also assumes that the first "+goal" in a rule removes this goal before adding—thus ensuring that models written for this ACT–R with one "+goal" run exactly as before.
2. The tuning and dialing models actually have one additional production rule that clears ACT–R's motor program when the task completes; this rule compensates for a problem with this driver model in which steering movements do not automatically clear the motor program as they would, given a complete motor module that integrates steering and other hand movements.

Acknowledgments

This work was supported by Office of Naval Research grant #N00014–03–1–0036. Many thanks to Frank Lee and Yelena Kushleyeva for guidance on the original model, and to John Anderson, Wayne Gray, Wai-Tat Fu, Mike Schoelles, Niels Taatgen, and Rob Goldstone for many comments and helpful suggestions.

References

Aasman, J. (1995). *Modelling driver behaviour in Soar.* Leidschendam, The Netherlands: KPN Research.

Altmann, E. M., & Gray, W. D. (2000). An integrated model of set shifting and maintenance. In N. Taatgen & J. Aasman (Eds.), *Proceedings of the 3rd International Conference on Cognitive Modeling* (pp. 17–24). Veenendaal, The Netherlands: Universal Press.

Altmann, E. M., & Gray, W. D. (2002). Forgetting to remember: The functional relationship of decay and interference. *Psychological Sciences, 13,* 27–33.

Altmann, E. M., & Trafton, J. G. (2002). Memory for goals: An activation-based model. *Cognitive Science, 26,* 39–83.

Anderson, J. R. (2005). Human symbol manipulation within an integrated cognitive architecture. *Cognitive Science, 29,* 313–341.

Anderson, J. R., Bothell, D., Byrne, M. D., Douglass, S., Lebiere, C., & Qin, Y. (2004). An integrated theory of the mind. *Psychological Review, 111,* 1036–1060.

Anderson, J. R., Bothell, D., Lebiere, C., & Matessa, M. (1998). An integrated theory of list memory. *Journal of Memory and Language, 38,* 341–380.

Anderson, J. R., & Lebiere, C. (1998). *The atomic components of thought.* Hillsdale, NJ: Lawrence Erlbaum Associates, Inc.

Anderson, J. R., Taatgen, N. A., & Byrne, M. D. (in press). Learning to achieve perfect time sharing: Architectural implications of Hazeltine, Teague, & Ivry (2002). *Journal of Experimental Psychology: Human Perception and Performance.*

Best, B. J., & Lebiere, C. (2003). Spatial plans, communication, and teamwork in synthetic MOUT agents. In *Proceedings of the 12th Conference on Behavior Representation in Modeling and Simulation.*

Byrne, M. D., & Anderson, J. R. (2001). Serial modules in parallel: The psychological refractory period and perfect time-sharing. *Psychological Review, 108,* 847–869.

Byrne, M. D., & Kirlik, A. (2005). Using computational cognitive modeling to diagnose possible sources of aviation error. *International Journal of Aviation Psychology, 15,* 135–155.

Card, S., Moran, T., & Newell, A. (1983). *The psychology of human–computer interaction.* Hillsdale, NJ: Lawrence Erlbaum Associates, Inc.

Chong, R. S. (1998). *Modeling dual-task performance improvement: Casting executive process knowledge acquisition as strategy refinement* (Tech. Rep. No. CSE-TR-378-98). Unpublished doctoral dissertation, Department of Computer Science and Engineering, University of Michigan, Ann Arbor.

Dingus, T. A., Antin, J. F., Hulse, M. C., & Wierwille, W. W. (1989). Attentional demand requirements of an automobile moving-map navigation system. *Transportation Research, A23,* 301–315.

Doane, S. M., & Sohn, Y. W. (2000). ADAPT: A predictive cognitive model of user visual attention and action planning. *User Modeling and User Adapted Interaction, 10,* 1–45.

Ellis, R. D., Goldberg, J. H., & Detweiler, M. C. (1996). Predicting age-related differences in visual information processing using a two-stage queuing model. *Journal of Gerontology: Psychological Sciences, 51,* 155–165.

Fick, C. S., & Byrne, M. D. (2003). Capture of visual attention by abrupt onsets: A model of contingent orienting. In F. Detje, D. Dörner, & H. Schaub (Eds.), *Proceedings of the Fifth International Conference on Cognitive Modeling* (pp. 81–86). Bamberg, Germany: Universitas-Verlag Bamberg.

Fincham, J. M., Carter, C. S., van Veen, V., Stenger, V. A., and Anderson, J. R. (2002). Neural mechanisms of planning: A computational analysis using event-related fMRI. *Proceedings of the National Academy of Sciences, 99,* 3346–3351.

Freed, M. (1998). Managing multiple tasks in complex, dynamic environments. In *Proceedings of the 1998 National Conference on Artificial Intelligence* (pp. 921–927). Menlo Park, CA: AAAI Press.

Fu, W.-T., Bothell, D., Douglass, S., Haimson, C., Sohn, M.-H, & Anderson, J. R. (2004). Learning from real-time over-the-shoulder instructions in a dynamic task. In *Proceedings of the Sixth International Conference on Cognitive Modeling* (pp. 100–105). Pittsburgh, PA: Carnegie Mellon University.

Gluck, K. A., Ball, J. T., Krusmark, M. A., Rodgers, S. M., & Purtee, M. D. (2003). A computational process model of basic aircraft maneuvering. In F. Detje, D. Doerner, & H. Schaub (Eds.), In *Proceedings of the Fifth International Conference on Cognitive Modeling* (pp. 117–122). Bamberg, Germany: Universitats-Verlag Bamberg.

Gluck, K. A., & Pew, R. W. (Eds.). (in press). *Modeling human behavior with integrated cognitive architectures: Comparison, evaluation, and validation.* Mahwah, NJ: Lawrence Erlbaum Associates, Inc.

Gray, W. D., & Schoelles, M. J. (2003). The nature and timing of interruptions in a complex cognitive task: Empirical data and computational cognitive models. In *Proceedings of the 25th Annual Meeting of the Cognitive Science Society* (p. 37). Mahwah, NJ: Lawrence Erlbaum Associates, Inc.

Hendricks, D. L., Freedman, M., Zador, P. L., & Fell, J. C. (2001). *The relative frequency of unsafe driving acts in serious traffic crashes.* Washington, DC: National Highway Traffic Safety Administration.

Hess, R. A., & Modjtahedzadeh, A. (1990). A control theoretic model of driver steering behavior. *IEEE Control Systems Magazine, 10*(5), 3–8.

Hildreth, E. C., Beusmans, J. M. H., Boer, E. R., & Royden, C. S. (2000). From vision to action: Experiments and models of steering control during driving. *Journal of Experimental Psychology: Human Perception and Performance, 26,* 1106–1132.

John, B. E. (1996). TYPIST: A theory of performance in skilled typing. *Human–Computer Interaction, 11,* 321–355.

Jones, R. M., Laird, J. E., Nielsen P. E., Coulter, K., Kenny, P., & Koss, F. (1999). Automated intelligent pilots for combat flight simulation. *AI Magazine, 20,* 27–42.

Just, M. A., Carpenter, P. A., & Varma, S. (1999). Computational modeling of high-level cognition and brain function. *Human Brain Mapping, 8,* 128–136.

Kelley, T. D., & Scribner, D. R. (2003). Developing a predictive model of dual task performance (Tech.Rep. ARL-MR-0556). Aberdeen, MD: U.S. Army Research Laboratory.

Kieras, D. E., & Meyer, D. E. (1997). A computational theory of executive cognitive processes and multiple-task performance: Part 1. Basic mechanisms. *Psychological Review, 104,* 3–65.

Kieras, D. E., & Meyer, D. E. (1997). An overview of the EPIC architecture for cognition and performance with application to human–computer interaction. *Human–Computer Interaction, 12,* 391–438.

Kieras, D. E., & Meyer, D. E. (1997). A computational theory of executive cognitive processes and multiple-task performance: Part 1. Basic mechanisms. *Psychological Review, 104,* 3–65.

Kieras, D. E., Meyer, D. E., Ballas, J. A., & Lauber, E. J. (2000). Modern computational perspectives on executive mental processes and cognitive control: Where to from here? In S. Monsell & J. Driver (Eds.), *Control of cognitive processes: Attention and performance XVIII* (pp. 681–712). Cambridge, MA: MIT Press.

Kieras, D. E., Wood, S. D., & Meyer, D. E. (1997). Predictive engineering models based on the EPIC architecture for a multimodal high-performance human–computer interaction task. *ACM Transactions on Computer–Human Interaction, 4,* 230–275.

Koechlin, E., Corrado, G., Pietrini, P., & Grafman, J. (2000). Dissociating the role of the medial and lateral anterior prefrontal cortex in human planning, *Proceedings of the National Academy of Sciences USA, 97,* 7651–7656.

Kushleyeva, Y., Salvucci, D. D., & Lee, F. J. (2005). Deciding when to switch tasks in time-critical multitasking. *Cognitive Systems Research, 6,* 41–49.

Laird, J. E., & Duchi, J. C. (2000). Creating human-like synthetic characters with multiple skill levels: A case study using the Soar Quakebot. In *Papers from the AAAI 2000 Fall Symposium on Simulating Human Agents* (Tech. Rep. FS-0A-03; pp. 75–79). Menlo Park, CA: AAAI Press.

Lallement, Y., & John, B. E. (1998). Cognitive architecture and modeling idiom: An examination of three models of the Wickens task. In M. A. Gernsbacher & S. J. Derry (Eds.), *Proceedings of the Twentieth Annual Conference of the Cognitive Science Society* (pp. 597–602). Hillsdale, NJ: Lawrence Erlbaum Associates, Inc.

Lebiere, C., Gray, R., Salvucci, D., & West, R. (2003). Choice and learning under uncertainty: A case study in baseball batting. In *Proceedings of the 25th Annual Conference of the Cognitive Science Society* (pp. 704–709). Mahwah, NJ: Lawrence Erlbaum Associates, Inc.

Lee, F. J., & Anderson, J. R. (2000). Modeling eye-movements of skilled performance in a dynamic task. In *Proceedings of the Third International Conference on Cognitive Modeling* (pp. 194–201). Gronigen: University of Groningen, Netherlands.

Liu, Y. (1996). Queueing network modeling of elementary mental processes. *Psychological Review, 103,* 116–136.

McClelland, J. L. (1979). On the time relations of mental processes: An examination of systems of processes in cascade. *Psychological Review, 86,* 287–330.

Meyer, D. E., & Kieras, D. E. (1997). A computational theory of executive cognitive processes and multiple-task performance: Part 2. Accounts of psychological refractory period phenomena. *Psychological Review, 104,* 749–791.

Miller, J. O. (1993). A queue-series model for reaction time, with discrete-stage and continuous-flow models as special cases. *Psychological Review, 100,* 702–715.

Nelson, R. (1995). *Probability, stochastic processes, and queueing theory—The mathematics of computer performance modeling.* New York: Springer Verlag.

Newell, A. (1990). *Unified theories of cognition.* Cambridge, MA: Harvard University Press.

Reed, M. P., & Green, P. A. (1999). Comparison of driving performance on-road and in a low-cost simulator using a concurrent telephone dialing task. *Ergonomics, 42,* 1015–1037.

Ritter, F. E., Baxter, G. D., Jones, G., & Young, R. M. (2000). Supporting cognitive models as users. *ACM Transactions on Computer Human Interaction, 7,* 141–173.

Salvucci, D. D. (2001a). An integrated model of eye movements and visual encoding. *Cognitive Systems Research, 1,* 201–220.

Salvucci, D. D. (2001b). Predicting the effects of in-car interface use on driver performance: An integrated model approach. *International Journal of Human–Computer Studies, 55,* 85–107.

Salvucci, D. D. (2002). Modeling driver distraction from cognitive tasks. In *Proceedings of the 24th Annual Conference of the Cognitive Science Society* (pp. 792–797). Mahwah, NJ: Lawrence Erlbaum Associates, Inc.

Salvucci, D. D. (in press). Modeling driver behavior in a cognitive architecture. *Human Factors.*

Salvucci, D. D., & Anderson, J. R. (2001). Integrating analogical mapping and general problem solving: The path-mapping theory. *Cognitive Science, 25,* 67–110.

Salvucci, D. D., Boer, E. R., & Liu, A. (2001). Toward an integrated model of driver behavior in a cognitive architecture. *Transportation Research Record, 1779,* 9–16.

Salvucci, D. D., & Gray, R. (2004). A two-point visual control model of steering. *Perception, 33,* 1233–1248.

Salvucci, D. D., Kushleyeva, Y., & Lee, F. J. (2004). Toward an ACT–R general executive for human multitasking. In *Proceedings of the Sixth International Conference on Cognitive Modeling* (pp. 267–272). Mahwah, NJ: Lawrence Erlbaum Associates, Inc.

Salvucci, D. D., & Macuga, K. L. (2002). Predicting the effects of cellular-phone dialing on driver performance. *Cognitive Systems Research, 3,* 95–102.

Schoppek, W. (2002). Examples, rules, and strategies in the control of dynamic systems. *Cognitive Science Quarterly, 2,* 63–92.

Smith, E. E., & Jonides, J. (1999, March 12). Storage and executive processes in the frontal lobes, *Science, 283,* 1657–1661.

Sodhi, M., Reimer, B., & Llamazares, I. (2002). Glance analysis of driver eye movements to evaluate distraction. *Behavior Research Methods, Instruments and Computing, 34,* 529–538.

Sohn, M-H., & Anderson, J. R. (2001). Task preparation and task repetition: Two-component model of task switching. *Journal of Experimental Psychology: General, 130,* 764–778.

Taatgen, N. (2005). Modeling parallelization and flexibility improvements in skill acquisition: From dual tasks to complex dynamic skills. *Cognitive Science, 29,* 421–455.

Taatgen, N. A., & Lee, F. J. (2003). Production compilation: A simple mechanism to model complex skill acquisition. *Human Factors, 45,* 61–76.

Taatgen, N., van Rijn, H., & Anderson, J. R. (2004). Time perception: Beyond simple interval estimation. In *Proceedings of the Sixth International Conference on Cognitive Modeling* (pp. 296–301). Pittsburgh, PA: Carnegie Mellon University.

Tsimhoni, O., & Liu, Y. (2003). Modeling steering using the Queuing Network—Model Human Processor (QN-MHP). In *Proceedings of the Human Factors and Ergonomics Society 47th Annual Meeting* (pp. 1875–1879). Santa Monica, CA: Human Factors and Ergonomics Society.

Vera, A., Howes, A., McCurdy, M., & Lewis, R. L. (2004). A constraint satisfaction approach to predicting skilled interactive performance. In *Human Factors in Computing Systems: CHI 2004 Conference Proceedings* (pp. 121–128). New York: ACM Press.

Wierwille, W. W. (1993). Visual and manual demands of in-car controls and displays. In B. Peacock & W. Karwowski (Eds.), *Automotive ergonomics* (pp. 229–320). London: Taylor & Francis.

Cognitive Science 29 (2005) 493–524

A Strategy-Based Interpretation of Stroop

Marsha C. Lovett

Department of Psychology, Carnegie Mellon University

Received 22 May 2004; received in revised form 18 January 2005; accepted 4 February 2005

Abstract

Most accounts of the Stroop effect (Stroop, 1935) emphasize its negative aspect, namely, that in particular situations, processing of an irrelevant stimulus dimension interferes with participants' performance of the instructed task. In contrast, this paper emphasizes the fact that, even with that interference, participants actually can (and usually do) exert enough control to perform the instructed task. An Adaptive Control of Thought–Rational (ACT–R) model of the Stroop task interprets this as a kind of learned strategic control. Specifically, the concept of utility is applied to the two processes that compete in the Stroop task, and a utility-learning mechanism serves to update the corresponding utility values according to experience and hence influence the competition. This model both accounts for various extant Stroop results and makes novel predictions about when people can reduce their susceptibility to Stroop interference. These predictions are tested in three experiments that involve a double-response variant of the Stroop task.

Keywords: Stroop; Strategy choice; Utility learning; Hybrid modeling; ACT–R; Computational modeling

1. Introduction

The Stroop effect (Stroop, 1935) is a long-studied, yet still intriguing, phenomenon in cognitive psychology. In its most general form, the Stroop effect occurs when two competing processes are relevant to the task at hand, but only one of these processes should govern the participants' response. The standard Stroop task requires naming the ink color of a word that, in some cases, spells a color (e.g., the word *red* printed in blue ink). The robust result is that performance varies as a function of the congruency between the ink color and the word. When the word spells a color that conflicts with the ink color, latencies and error rates increase relative to nonword and non-color-word controls, an effect known as Stroop *interference.* Conversely, Stroop *facilitation* occurs when the word spells a color that matches the ink color, with laten-

---

Requests for reprints should be sent to Marsha C. Lovett, Department of Psychology, Carnegie Mellon University, 124 Cyert Hall, Pittsburgh, PA 15213–3890. E-mail: lovett@cmu.edu

cies and error rates decreasing relative to controls. The most important aspect of Stroop phenomena is that these effects disappear under word-reading instructions.

Theories to explain Stroop effects abound. Each tries to explain people's *lack* of cognitive control, their inability to ignore the word when responding to the color. One explanation is that, through a lifetime of practice, reading has become so automatized that it impacts processing, even under color-naming instructions (Cohen, Dunbar, & McClelland, 1990). Other theories include response compatibility (Dalrymple-Alford & Azkoul, 1972), differential translation requirements (Virzi & Egeth, 1985), and speed of processing (Schooler, Neumann, Caplan, & Roberts, 1997). Evaluating these explanations often involves identifying and testing qualitative predictions they make. For example, a pure horse-race model has been discounted because results failed to support its prediction of a reversed Stroop effect when the ink color sufficiently precedes the word (Glaser & Glaser, 1982). Similar manipulations of stimulus onset asynchrony have been used to argue for modifications to a pure translation model of the Stroop effect (Sugg, & McDonald, 1994).

More recently, computational modeling has been especially productive in refining our understanding of the Stroop effect. First, quantitative predictions of performance, such as the relative size of particular Stroop effects, can more sharply distinguish among theories, especially in cases where qualitative differences between theories tend not to arise. Second, explanations of the Stroop effect often invoke theoretical constructs that require careful operationalization, which computational models by their very nature produce. Third, in the context of a cognitive architecture (cf. Newell, 1990)—where a set of fixed computational mechanisms applies *across* models—results from Stroop studies can constrain basic processes that are exercised in a broad range of other tasks.

1.1. Many effects, many models

The basic Stroop experiment has been varied in numerous ways, producing a diverse set of empirical effects that offer useful constraints for models of Stroop. In his 1991 review, MacLeod identified a list of Stroop-related findings that "must be captured by any successful theory [or computational model] of the Stroop effect" (p. 163). Table 1 presents a subset of that list, highlighting particularly influential results for testing computational accounts of the Stroop effect.

Given such a powerful set of modeling constraints, it has been a challenge to develop a model that meets all of them. Nevertheless, there have been many successful models fit to interesting subsets of the results (e.g., Altmann & Davidson, 2001; Botvinick, Braver, Barch, Carter, & Cohen, 2001; Cohen et al., 1990; Cohen & Huston, 1994; Phaf, van der Heijden, & Hudson, 1990; Roelofs, 2000, 2003; Roelofs & Hagoort, 2002). Three examples illustrate the range of computational approaches that have been taken to account for Stroop phenomena—two connectionist models (Cohen et al., 1990; Phaf et al., 1990) and a production-system model (Roelofs, 2000).

The Cohen et al. (1990) model and its extensions (e.g., Cohen & Huston, 1994; Botvinick et al., 2001) represent the two competing processes as separate pathways in a connectionist network. The competition between pathways is managed by task-control nodes that "gate" information processing and create a bias toward the instructed task. The central concept in this

Table 1
Empirical results that constrain models of the Stroop effect

Description of the Effect	Representative Citation(s)
The basic effect: The basic effect is robust to methodological variations, including list versus single-trial presentation, and task variants, such as the picture-word task. (1)	Dalrymple-Alford and Budayr, 1966
Semantic gradient: For noncolor words, the size of the Stroop effect increases with the strength of the semantic association to a color concept. (3)	Dalrymple-Alford, 1972
Facilitation: Facilitation can occur on congruent trials, but the size of this effect is smaller than interference and depends on the choice of neutral trials. (5)	various
Proportion of trial types: The proportion of trials of different types (conflict, congruent, and neutral) impacts the size of the Stroop effect. (7)	Tzelgov , Henik, and Berger, 1992
SOA: The maximal Stroop effect occurs when the color and word components of the stimulus are presented within 100 msec of each other. When the color precedes the word, a reverse Stroop effect (i.e., interference in the word-reading task) is *not* found. (10 & 11)	Glaser and Glaser, 1982
Degree of practice: The degree of practice at processing each of the two stimulus dimensions influences which task will interfere with the other and to what degree. (12)	MacLeod and Dunbar, 1988
Response modality: The modality of response matters with larger Stroop effects for oral rather than manual responses; also response compatibility impacts the size of Stroop effects. (13)	various

Note. In parentheses after each effect's description is the corresponding effect number from MacLeod's (1991) list of 18 major results. SOA = Stimulus-Onset Asynchrony.

model, *graded automaticity,* posits that greater practice at a task produces greater automaticity. The more automatized a task is relative to its competitor, the more its processing can proceed without additional input. Hence, word reading interferes with color naming even when the instructed task is color naming.

The Phaf et al. (1990) model was built as a Stroop extension to a connectionist model of visual attention. This model's architecture differs qualitatively between word reading and color naming in that there are direct input–output connections for word reading but not for color naming. This word-reading shortcut reduces the interference from color naming on word reading but allows for interference in the other direction.

The Roelofs (2000, 2003; Roelofs & Hagoort, 2002) model of Stroop phenomena was built as an extension to the WEAVER++ model of word production (Levelt, Roelofs, & Meyer, 1999) and specifies several stages of processing for word reading. Like the Phaf et al. (1990) model, it establishes a word-reading advantage by requiring fewer processing steps for that task, enabling different-sized congruency effects between word reading and color naming. This model was elaborated by Altmann and Davidson (2001) to include several aspects of Adaptive Character of Thought–Rational (ACT–R)'s declarative memory system.

1.2. *A place for strategy in the Stroop task*

What is immediately striking about the Stroop effect is that word reading *interferes* with color naming. And yet, despite this interference, participants answer the vast majority of trials correctly. Error rates range from 2% to 10% when the task involves naming the color of the ink. This implies that there is a mechanism governing the choice between the two processing pathways that takes into account the instructed goal. In the problem-solving literature, this is called *strategy choice* (e.g., Lovett, 1998; Siegler, 1991, 1996). In the context of the standard Stroop task, the word *strategy* refers to the procedure for processing the word or the ink color. It is worth noting that the word *choice* here need not, and usually does not, imply a decision made with conscious awareness. Although previous work on the role of strategy in Stroop (e.g., Cheesman & Merikle, 1984; Logan & Zbrodoff, 1982; Logan, Zbrodoff, & Williamson, 1984; Neill, 1978), has focused on explicit strategic approaches, this article focuses on the implicit strategy choice between word reading and color naming.

The literature on strategy choice in problem solving reveals that it is not only how much a strategy has been practiced that impacts its processing but how effective and efficient it was when applied (e.g., Lovett & Anderson, 1996; Siegler, 1996). One way to quantify "effectiveness and efficiency" is through a strategy's *utility*. This is a function of the strategy's expected gain—its likelihood of leading to success multiplied by the value attributed to that success—minus the expected costs of getting there. With *utility learning,* each strategy's associated utility value is updated based on its success and cost of achieving success when it is applied. A strategy with greater utility relative to its competitors will be more likely to be chosen in future situations where it is relevant. This utility-learning view includes practice as a part of effective strategy choice but further specifies that, in order for a strategy to become a more prepotent response, its applications must have led to high-success–low-cost outcomes—that is, high-utility outcomes.

Both a utility-learning model and a practice-based model, such as Cohen et al. (1990), specify a general mechanism for learning by experience that applies to diverse tasks and phenomena. Both predict that Stroop effects can occur between competing processes beyond word reading and color naming. Both predict that giving more practice to one of the competing processes will lead to Stroop effects. In the practice-based model, this is a direct effect of relative strength: The more practiced process will be more automatized and hence more prepotent. In a utility-learning model, extra practice tends to confer a utility advantage (e.g., practice leads to speedup, which reduces cost), leading to the same result. But a utility-learning model makes qualitatively different predictions from a purely practice-based model when the utility of the competing processes differs while the amount of practice is held constant. In this case, a utility-learning model predicts a Stroop effect, whereas a practice-based model does not.

This article describes a strategy-based, utility-learning model of Stroop and shows how it accounts for the results in Table 1 as well as new empirical findings that bear out the prediction just mentioned. The following section provides a description of this model and explains how it accounts for various extant Stroop results. Next, the article describes an infrequently studied variant of the Stroop task, called *double-response Stroop,* where participants are asked to respond to *both* aspects of the stimuli. Double-response Stroop is particularly interesting from a strategy-choice perspective because participants' relative preference for color-naming or word-reading can be observed on a trial-to-trial basis, for example, by observing which of the two stimulus dimensions is reported first. This double-response task is used to test this strategy- and utility-based model in a series of three experiments.

2. A strategy-choice model of Stroop

Most models of Stroop specify "universal" processing for each trial type in that the same steps or cycles are engaged regardless of the system's state (cf. Siegler, 1996). Although there is stochasticity in these models' processing (e.g., different responses and response latencies can be produced), the order in which information is processed and the kind of information that is processed is the same for every trial. In contrast, modeling Stroop effects as the result of strategy learning and choice—what this article calls a *strategy-based* perspective on Stroop—leads to a variety of ways that information can be processed; for example, different trials could focus on different kinds of information or process the same information but in different sequences. The difference between strategy-based and universal processing is akin to the difference between "fixed effects" statistical models, where the effects have fixed but unknown values, and "random effects" statistical models, where the effects are considered to be drawn from a population. Just as there is another layer of noise in "random effects" statistical models, so there is another level of (systematic) variation in strategy-based computational models. This extra variation in strategy-based models leads to the prediction of qualitative (not just quantitative) variation between individuals and even between individual trials completed by the same person. There are many examples of individual differences in when and whether Stroop effects occur (e.g., Chrisman, 2001; Comalli, Wapner, & Werner, 1962; Kane & Engle, 2003; Schiller, 1966). As for systematic trial-to-trial differences within individual participants, neuroimaging data is one source of evidence (Kerns et al., 2004).

Figure 1 presents a diagram of the different processing paths that this strategy-based model takes. This diagram highlights three branching points. One occurs early and diverts processing to focus on either the word or color dimension of the stimulus. Then, along both of these paths, there is another branching point that involves either immediately responding to the processed dimension or first checking that the task-relevant dimension has been processed. In addition, although it is not highlighted in this diagram, there is yet another opportunity for processing differences after task checking on both paths: When task checking reveals that the processed dimension matches the instructions, the model generates a response directly; otherwise, the model continues processing the stimulus, now focusing on the alternate dimension. This model allows for six qualitatively distinct processing pathways. Note that four of these involve task checking, which is quite likely to produce the correct response but takes extra time. The other two (depicted at the left and right extremes in Figure 1) save time by not checking but may respond with the task-irrelevant dimension—an error on conflict trials.

Besides supporting these qualitative predictions about strategy variation in Stroop, this model makes quantitative predictions that depend on the specific computational mechanisms driving the execution and learning of the model's strategies. The model represents strategies as *production rules*—contingencies for action of the form "When <conditions> are true, then do

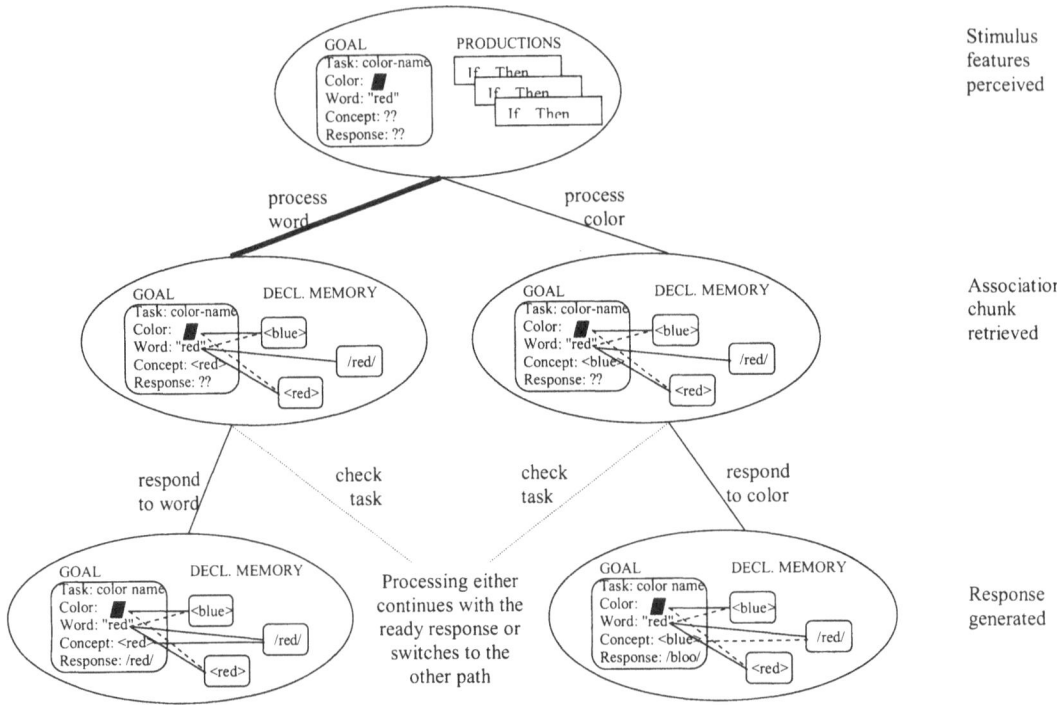

Fig. 1. Flow of control in this model for a color-naming trial. Lines between ovals represent production rules (thicker lines indicate greater likelihood of firing), and ovals represent parallel retrievals that modify information in the current goal. Along both the word-reading and color-naming paths, there is an optional check before responding (two lines labeled "check task").

<actions>." In this model, the color-naming production rule specifies as its condition that the task must be color naming. In contrast, the word-reading production rule has a very general condition; it will fire whenever the stimulus has wordlike features. So, when the task instructions specify color naming, both options match. However, the model only executes one production rule at a time, and that choice is governed by the utility values associated with the production rules: Generally speaking, the higher a production rule's utility value relative to its competitors, the more likely it is to be executed. (In Figure 1, this is represented by different line thickness of the production rule lines.) The model's computational mechanisms for using and updating utility values (discussed in the next subsection) enable specific quantitative predictions about how likely the model's different processing paths are under various circumstances. For example, these mechanisms enable predictions about how likely it is that, say, a word response will be given without task checking when the task is color naming and there were 70% conflict trials.

When the model executes a particular production rule, it performs the actions associated with that production rule. Often this involves modifying the current goal based on information retrieved from memory (see ovals in Figure 1). For the process-word production rule, the goal is updated to reflect what is retrieved about the word dimension of the stimulus. That retrieval process involves a parallel competition among *word-association chunks,* so called because they link information from a word (the stimulus *red*) to its related concept (the concept of *redness*, or <red>), and possibly even to the associated motor program for reading that word (the verbal response /red/). For the process-color production rule, a similar action is performed, but there the retrieval involves a parallel retrieval of *color*-association chunks, each of which links a perceived ink color to the corresponding color concept. Whereas the competition among production rules is based on production rules' utility, the retrieval of chunks is based on a quantity called *activation,* which reflects how likely a chunk is to be needed, given its past use. The model's computational mechanisms for using and updating activation values (see next subsection) enable specific quantitative predictions about how likely different chunks are to be retrieved and how long each retrieval will take.

2.1. ACT–R implementation

The basic knowledge structures in this model are (a) production rules representing the various strategies participants employ in the Stroop task and (b) chunks representing the various facts and relations that these strategies require. The model is implemented in the cognitive architecture ACT–R (Anderson & Lebiere, 1998), which specifies a fixed set of mechanisms governing how such structures are used. In any ACT–R model, these same mechanisms are applied to derive quantitative predictions about learning and performance. In this Stroop model, three of ACT–R's mechanisms are particularly important: utility-based choice, utility learning, and activation-based retrieval. Each of these will be discussed in turn.

In each cycle of ACT–R's processing, a single production rule, from the set of those matching this situation, fires, and its actions are executed. ACT–R's utility-based choice mechanism specifies that the production rule with the highest utility, after some noise is added, is the one to fire. Specifically, the probability that production rule i fires depends on its utility (U_i), the utility of its competitors (U_j), and the distribution of the added noise (here, logistic noise with variance t):

$$P(i\ fires) = \frac{e^{-U_i/t}}{\sum_j e^{-U_j/t}}$$

In this model, this choice applies to each branching point in Figure 1. For the first choice, the process-word production rule has a higher utility than process-color, so the model is more likely to process the word dimension of the stimulus first. However, because the choice process is noisy, this will not always occur, producing qualitatively different trials, that is, those initiated by word reading versus color naming. Note that relative production-rule utility can be viewed as this model's operationalization of graded automaticity of processing: The higher one production rule's utility value is relative to another, the more likely it will dominate in the choice competition.

But where do these utility values come from? In ACT–R, utility of production rule i (U_i) is defined as $U_i = P_i * G - C_i$, where P_i is the estimated probability that production rule i leads to success, G is the value attributed to achieving success, and C_i is the estimated cost of achieving success (measured in units of time to complete this goal). Each time this goal is completed,

ACT–R automatically updates the utility value for each of the production rules that fired as a part of that goal completion. This involves updating the two components of utility, P and C, for each relevant production rule i at that point in time t:

$$P_i(t) = [N_i(t-1)*P_i(t-1) + \text{CurrentSuccess}]/N_i(t)$$

$$C_i(t) = [N_i(t-1)*C_i(t-1) + \text{CurrentCost}_i]/N_i(t)$$

Here, $N_i(t-1)$ is the number of times production rule i had been fired up to time $t-1$, CurrentSuccess is an indicator variable for whether this goal was completed successfully, and CurrentCost$_i$ is the time between production rule i firing and goal completion, that is, the time to achieve success for that production rule). Combining these utility-updating equations with ACT–R's choice mechanism implies that the more successful and less costly a production rule is in practice, the higher its utility value and hence the more likely it will fire. In general, these two mechanisms allow ACT–R models to learn to prefer strategies that are more effective and efficient. In this model of Stroop, if there were some kind of practice that would enable color naming to become more successful and less costly (or word reading to become less successful and more costly), the relative utility values would shift, thus increasing the likelihood that color naming fires, and reducing the prepotency of word reading. We test this prediction in Experiments 2 and 3.

The third ACT–R mechanism, activation-based retrieval, specifies the time to retrieve a chunk i as $F*\exp(-A_i)$, where A_i is the chunk's total activation and F is a latency-scaling parameter. But what is total activation? It is the sum of *base-level activation* and *source activation*. A chunk's base-level activation increases with practice and decreases with delay according to ACT–R's declarative learning mechanism (cf. Anderson & Lebiere, 1998). A chunk's source activation reflects its relevance to this goal and is computed via the network of links connecting chunks (positively and negatively) to each other. Source activation spreads from the goal along these links, giving extra activation boosts (or dips) to chunks that are relevant (or irrelevant) to this goal. Thus, total activation of chunk i is

$$A_i = B_i + \Sigma W_j S_{ji}$$

where B_i is the base-level activation of chunk i, W_j is the amount of activation spreading from component j of this goal, and S_{ji} is the $i{\rightarrow}j$ link strength.

In this model, base-level activation for word-association chunks is preset to be higher than that for color-association chunks (2 vs. 0), reflecting greater prior practice at retrieving word-related information. In addition, the model was given the following fixed link strengths: (a) +1.5 for each pair of chunks involving matching colors, (b) –1.5 for each pair of chunks involving mismatching colors, (c) +0.6 for each pair of chunks involving matching tasks, (d) –0.6 for each pair of chunks involving mismatching tasks, and (e) 0 for all other pairs. Because the stimulus is represented in the goal, source activation spreads from the two stimulus dimensions along these links to various chunks in memory. In conflict (congruent) trials, this reduces (increases) the target chunk's total activation because of negative (positive) link strengths just mentioned.

2.2. How does the model account for basic Stroop?

The model's production-rule choice mechanism favors word reading (even when the task is color naming) because the process-word production rule has high initial utility. This means that the word will likely be processed (at least to some degree) first. Because word-association chunks have high base-level activations, the time required for their retrieval will be short. Moreover, the source activation increase (decrease) from congruent (conflict) trials will have little effect on retrieval time, because the retrieval time function is nonlinear. In some cases, the model will simply respond at this point, making word-reading trials highly accurate and fast, with little effect of trial congruency.

If the task is color naming, however, responding at this point will produce an error for conflict trials and a success for congruent trials. Thus, the model predicts more errors on conflict trials for the color-naming task. If, instead of responding immediately, the check-task production rule fires, a second pass of processing will be initiated. The relevant color-association chunk will be retrieved with a latency determined by its total activation. For congruent trials, total activation is boosted from source activation spreading from the stimulus color, the matching word, the current task, and the previously retrieved, matching word-concept. For conflict trials, source activation contributions to total activation are mixed: There are additions from the stimulus color and current task but reductions from the mismatching word and word concept. Because color-association chunks have moderate base-level activations and because the latency function is nonlinear, source activation contributions have a significant impact on color associations' retrieval times. Thus, for color naming, the model responds faster for congruent trials than for conflict trials, producing Stroop facilitation and interference.

It is worth noting that this model is consistent with a recent view that Stroop facilitation reflects participants sometimes making word-based responses on congruent, color-naming trials (MacLeod & MacDonald, 2000). Related work highlights the fact that, empirically speaking, Stroop facilitation is not the exact inverse of Stroop interference (e.g., MacLeod, 1998). In this model, Stroop facilitation stems from two influences—source activation boosts to memory retrieval and the possibility of responding quickly with the irrelevant dimension—the second of which does not have its inverse in Stroop interference.

2.3. Basic model fits

Besides explaining the model's functioning in the standard case, it is important to demonstrate that the model fits key results in the Stroop literature. Lovett (2002) showed that a preliminary version of this model accounted for five experiments containing 92 data points. This fit was performed over all 92 data points by varying 12 parameters; it produced an R^2 of .95 and mean deviation of 33.4 msec. The 12 varied parameters included a separate latency factor F and intercept for each of the five experiments plus an extra free parameter in two cases where atypical stimuli were used. These five experiments correspond to rows 1, 2, 5, and 6 from Table 1 plus an additional study indicating magnified Stroop interference among schizophrenic patients.

Here, we present the model's account for one experiment from that set—Macleod and Dunbar (1988)—because it is most relevant to strategy learning. In this experiment, participants learned a shape-naming task where different shapes were assigned color-word names. Over the course of

(a)   (b)

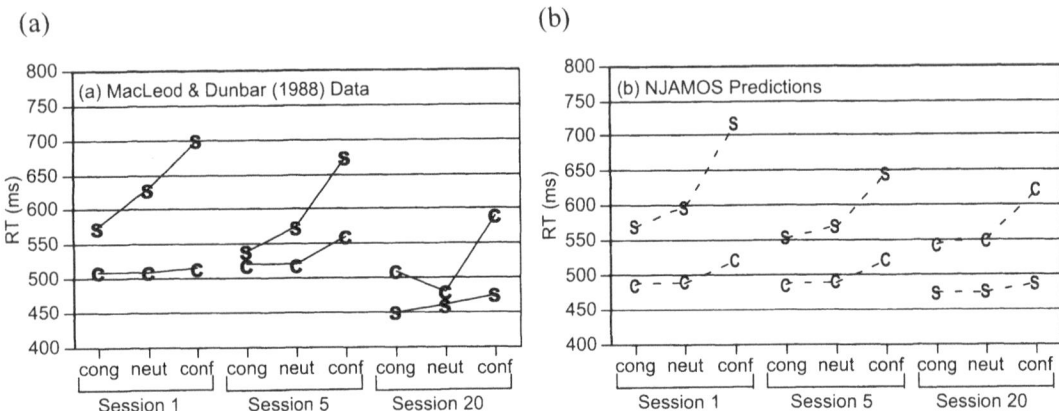

Fig. 2. Data from MacLeod and Dunbar (1988) (a) and this model's predictions (b). In each panel, RTs for shape naming (S) and color naming (C) are plotted separately across congruency and training manipulations.

20 days of practice at the shape-naming task, participants were tested intermittently in a Stroop task variant where shape naming and color naming compete (i.e., the shapes were presented in different ink colors to create congruent, neutral, or conflict trials under either shape-naming or color-naming instructions). From the data in Figure 2a, one can see that, at Day 1, color naming interfered with shape naming but not vice versa. At Day 20, these effects were reversed, and at Day 5 the results showed moderate interference in both directions.

Figure 2b presents our model's fit to these data. Note that this is a case where, besides the latency factor F and a general intercept, an additional parameter was used to represent the initial base-level activation for shape association. From this initial value, ACT–R's declarative chunk-learning mechanism was applied to update the base-level activations as the model received additional practice at shape naming during training. Note that increases in base-level activation make the shape chunks faster to be retrieved and less susceptible to trial-type effects. Also, over the course of training, the model updates the utilities of the shape-processing and color-processing production rules based on their effectiveness and efficiency in use. The change is quite dramatic for the shape-processing production rule because this is a completely novel task, and there is a lot of room for utility improvement. Specifically, the process-shape production rule's utility rises as it gets faster (because of the chunk learning mentioned earlier). This, in turn, makes the model more likely to process the shape first (because of utility-based choice). This strategy shift is another means by which the model speeds up in its shape naming across training; the model is no longer starting the shape naming after some initial color processing. Moreover, processing the shape first increases the size of the trial-type effect for color naming because the retrieved shape information spreads source activation to the relevant color-association chunk—a positive contribution on congruent trials and negative contribution on conflict trials.

Although MacLeod and Dunbar's (1988) experiment showed that practice at a nondominant task in a Stroop-like competition can reverse initial Stroop effects, practicing the standard Stroop task with equal proportions of the three trial types generally does *not* lead to a sizable reduction in the Stroop effect. This suggests that to produce noticeable change in participants' control of processing in the Stroop task, a more powerful impetus to learning must be established. This was accomplished via a new skill in MacLeod and Dunbar and has also been accomplished by increasing the proportions of conflict trials in various other studies (Cheesman & Merikle, 1984; Lindsay & Jacoby, 1994; Logan, 1980; Logan & Zbrodoff, 1979; Tzelgov, Henik, & Berger, 1992). In these studies, the general result is as follows: The more frequent the conflict trials, the smaller the size of the interference effect. Indeed, Experiment 3 from Stroop's original article gave participants repeated sessions of 100% conflict trials and produced both a striking reduction in the Stroop effect and even a reverse effect, that is, color naming interfered with word reading.

Although fits are not presented here, this model can account for such results nicely without further elaboration. The model simply updates the relative utility values of color naming and word reading through experience. The utility of color naming starts lower than word reading, but with a higher frequency of conflict trials, the utility value of color naming rises more. This leads to a higher probability of processing color information first under color-naming instructions and hence lowers the size of Stroop interference.

2.4. Logic of the experiments, task variant, and manipulations

Like MacLeod and Dunbar's (1988) shape-naming task and the manipulation of conflict trials' frequency, the experiments described in this article present participants with a more-difficult-than-usual Stroop situation, with the intention of impacting Stroop interference by producing a greater impact on the competing strategies' utility. Specifically, the following three experiments make use of the double-response Stroop task (Greenwald, 1972; Klein, 1964; Schweickert, 1978; Shimada & Nakajima, 1991) in which participants are shown standard Stroop trials but asked to respond to *both* stimulus dimensions. This task was chosen not only for its potential to produce larger utility changes but because it is particularly informative to a strategy-based view of Stroop, in that one can observe whether a participant responds to the word first or the color first and get a window onto potential strategy choices.

Past research on the double-response Stroop task has focused on the effects of response orders and response-modality mappings. Klein (1964) found that a word-first response order was easier than a color-first response order, but Shimada & Nakajima (1991) found no such difference. (Because both studies used 100% conflict trials, there is no assessment of the relative size of Stroop effects for the two response orders.) Greenwald (1972) and Schweickert (1978) found that a high-compatibility response-mapping (press button for color, say word) was easier than a low-compatibility mapping (press button for word, say color) and that this effect interacted with Stroop interference.

The double-response task experiments presented here touch on some of the same issues but focus on implications of our model of Stroop. In particular, the first experiment manipulates response-modality mapping (similarly to Greenwald, 1972, and Schweickert, 1978) but does so in a context where participants' response orders are not constrained. This enables a test of whether there is systematic strategic variation in response order, as a strategy-based view of Stroop would predict. The second and third experiments manipulate response order (similarly to Klein, 1964, and Shimada & Nakajima, 1991) but do so in a context where Stroop interference can be measured. This enables tests of specific predictions about change in Stroop interference that stem from the utility-based learning mechanism employed in this strategy-based model of Stroop.

3. Experiment 1

The goal of Experiment 1 was to explore what strategies participants naturally generate and exercise in the Stroop task. With the double-response task, one can measure, on a trial-by-trial basis, whether a participant responds to the word first or the color first. Moreover, if strategic differences are observed between individuals (or across trials within individuals), this experiment can test whether those differences are related to Stroop interference. If so, this would provide preliminary evidence that strategies do play a role in modulating Stroop interference, both directly (as measured in the double-response blocks) and indirectly (as measured in the standard Stroop blocks).

3.1. Methods

3.1.1. Participants

The participants in this experiment were 36 Carnegie Mellon University undergraduates who received course-related credit for participating. Participants were screened to confirm that they were right-handed, English speakers with full-color vision.

3.1.2. Design

The design of this experiment included one between-subject factor, namely whether the participants were instructed to respond to the color dimension manually and the word dimension verbally or vice versa. Seventeen participants were randomly assigned to the condition "Key in the color and say the word," and 19 participants were randomly assigned to the condition "Say the color and key in the word." As in the Greenwald (1972) and Schweickert (1978) studies of the double-response Stroop task, responding to the color manually and the word verbally can be considered the more natural or *high-compatibility* mapping, whereas responding to the color verbally and the word manually can be considered the less natural or *low-compatibility* mapping. Note that in *neither* condition were participants instructed as to the order in which they should make their responses in the double-response task. Instead, they were told (a) to respond as quickly as they could while maintaining accuracy and (b) to make each response as soon as they were able (i.e., to avoid "bundling" their responses).

The other factors included in this experiment involved the within-subject manipulations of task type and trial type. Specifically, there were four different tasks—color-only, word-only, standard Stroop, and double-response Stroop, each organized into blocks of 48 trials. The sequence of blocks was presented as three superblocks, each composed of the following: one-color-only block, one-word-only block, one standard Stroop block, and three double-response Stroop blocks. That is, participants completed 18 blocks (3 superblocks of 6 blocks each) at 48 trials each, for a total of 864 trials. Blocks were randomly ordered within each superblock for each participant. For the standard Stroop and double-response Stroop blocks, there were equal numbers of conflict and congruent trials.

3.1.3. Procedure

After a brief questionnaire to confirm eligibility and to collect demographic data, participants were given a brief overview of the study. Specific task instructions were presented on the computer screen using E-Prime software (Psychology Software Tools, Pittsburgh, Pennsylvania), and some of these were read aloud by the experimenter for emphasis. For example, participants were instructed differently, depending on their condition, as to how they should respond using the button box and by speaking into a voice-activated microphone. After these instructions were read and any questions answered, the experimental trials began. A task-specific instruction screen preceded each involving a new task.

3.1.4. Analysis

Reaction-time data were analyzed by computing the median response time for each participant over the relevant set of trials (e.g., a given block), for correct trials only. These medians were submitted to subsequent analyses, with aggregate results typically presented as means of these medians. Accuracy data were computed as the mean error rate for each participant over the appropriate set of trials. In all analyses, statistical significance was calculated based on an alpha level of 0.05. Note that for Stroop interference effects, the difference between conflict and congruent trials is used because this design did not include any neutral trials.

3.2. Results and discussion

The main findings of this experiment are (a) that participants *did* vary in their chosen order of response in the double-response task and (b) that these self-selected participant groups exhibited significantly different Stroop interference in the *standard* Stroop task. In this section we will describe these and other basic results in detail, organized by task type.

Figure 3 presents the reaction-time data for both response-mapping conditions and all four task types (with conflict and congruent trials presented separately for the two Stroop tasks). Corresponding accuracy data were very high—above 97% everywhere except for the double-response conflict trials in the say color-key word condition, where accuracy was 95%.

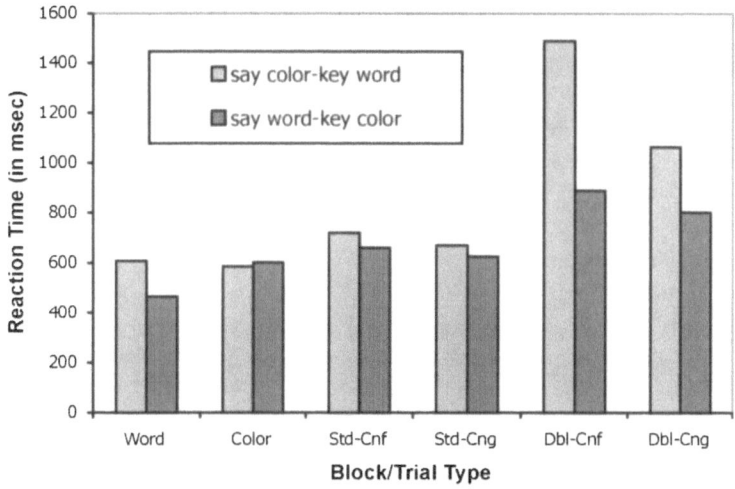

Fig. 3. Experiment 1 reaction-time data for both conditions and all four block types (conflict and congruent trials treated separately).

Thus, analyses of variance (ANOVAs) were conducted on the reaction-time data only. These analyses investigated, for each task, the effects of response-mapping condition, repetition (across three or nine blocks, depending on the task), trial type (for standard and double-response Stroop tasks only), and their interactions. The columns in Table 2 present the results of these analyses by task type, whereas the rows summarize whether each factor (or interaction) was significant across task types.

As the first row of Table 2 indicates, there was a difference in reaction time between the two conditions for all tasks except the color-only task. Not surprisingly, in the three tasks showing this effect, the high-compatibility condition responded significantly faster. The absence of this effect for the color-only tasks suggests that—regardless of response mapping—participants were consistently slow to respond to the ink color, even when it was the only stimulus dimension that they were either saying or keying.

Referring to the second row in Table 2, all four tasks showed a main effect of block repetition. For the color-only and double-response Stroop tasks, this main effect reflects a speedup across blocks. For the word-only and standard Stroop tasks, there was a speedup between Blocks 1 and 2 but then an increase in reaction times for Block 3. It is unclear why these reaction times would rise in Block 3 after a predictable decrease from Block 1 to Block 2, unless perhaps fatigue set in. It is also worth noting that block repetition did not show a significant interaction with condition for any of the tasks (see Row 3 of Table 2).

For the standard and double-response Stroop tasks, there are additional results involving trial type (conflict vs. congruent). Not surprisingly, the main effect of trial type was significant in both tasks, with conflict trials taking longer. This reflects the presence of Stroop interference

Table 2
ANOVA results for Experiment 1

| Factor | Task Type | | | |
	Word-Only	Color-Only	Standard Stroop	Double-Response
Response Mapping	$F(1, 34) = 63.9^{**}$, $MSE = 8682$	$F(1, 34) = 0.49$ $MSE = 14152$	$F(1, 34) = 3.9^{*}$, $MSE = 35900$	$F(1, 34) = 77.5^{***}$, $MSE = 400400$
Block Repetition	$F(2, 68) = 19.4^{**}$, $MSE = 1213$	$F(2, 68) = 3.9^{*}$, $MSE = 1426$	$F(2, 68) = 12.0^{**}$, $MSE = 4280$	$F(8, 272) = 60.7^{***}$, $MSE = 15566$
RM × BR	$F(2, 68) = 1.7$, $MSE = 1213$	$F(2, 68) = 1.7$, $MSE = 1426$	$F(2, 68) = 0.9$, $MSE = 4280$	$F(8, 272) = 0.7$, $MSE = 15566$
Trial Type	n/a	n/a	$F(1, 34) = 45.4^{***}$, $MSE = 2215$	$F(1, 34) = 91.3^{***}$, $MSE = 117600$
TT × RM	n/a	n/a	$F(1, 34) = 1.8$, $MSE = 2215$	$F(1, 34) = 39.5^{***}$, $MSE = 117600$
TT × BR	n/a	n/a	$F(2, 68) = 22.5^{***}$, $MSE = 1794$	$F(8, 272) = 4.6^{***}$, $MSE = 7000$
TT × RM × BR	n/a	n/a	$F(2, 68) = 5.4^{**}$, $MSE = 1794$	$F(8, 272) = 2.4^{*}$, $MSE = 7000$

Note. ANOVA = analysis of variance; MSE = mean square error; RM = Response Mapping; BR = Block Repetition; TT = Trial Type; n/a = not applicable.

$^{*}p < .05.$ $^{**}p < .01.$ $^{***}p < .001.$

in both tasks. It is interesting to note that, for the double-response Stroop task, there was also a significant interaction between trial type and condition. This interaction stems from the size of the Stroop interference effect being larger in the low-compatibility condition. This is not very surprising if we consider that the size of Stroop interference is simply "scaling" with the overall reaction time, which is so much larger for the low-compatibility participants in the double-response task. Also, this result is consistent with the findings of Greenwald (1972) and Schweickert (1978).

Other effects in the standard and double-response Stroop tasks involve the interaction between trial type and block repetition and the three-way interaction of Trial type × Block × Condition. The Trial type × Block interaction is significant for both Stroop tasks and likely reflects a general shrinking of the trial-type effect across blocks, which would be another example of Stroop interference scaling with overall reaction time. Finally, the three-way interaction is significant for both Stroop tasks. Although this high-order interaction is by its nature complex, the simplest interpretation is that the size of the trial-type effect (i.e., Stroop interference) scales with the overall reaction time, which is much greater for the low-compatibility condition and also much greater in the earlier blocks of the experiment.

Drawing firm conclusions about these Stroop interference effects, however, may not be warranted given the result foreshadowed earlier, namely, that Stroop interference effects were different for groups of participants who naturally exhibited different response orders in the double-response Stroop task. The next subsection describes this form of strategic variation, as observed in the double-response task, and presents a strategy-based analysis of Stroop interference effects.

3.2.1. Different strategies in the double-response Stroop task

The strategy analysis conducted on the double-response Stroop data involved assessing, for each double-response trial, whether the participant responded with the color dimension or the word dimension first. Recall that participants had free choice to respond word-first or color-first on a trial-by-trial basis. Figure 4 shows, for 6 selected participants in the high-compatibility condition, the proportion of word-first trials in each double-response block. There are several features of these data worth noting. First, the vast majority of the participants—31 out of 36 across both conditions (4 out of the 6 participants shown)—responded predominantly in a single order; that is, more than 90% of the time word-first or more than 90% of the time color-first. This is a striking result because there is no logical constraint of the task to choose either response order nor to *maintain* a previously chosen response order. Second, within the low-compatibility response-mapping condition (not depicted in Figure 4), the vast majority of participants (15 out of 17) were categorized as responding predominantly color-first. For these participants, color was reported verbally, so this result may reflect that the verbal response mode is either faster or more dominant. Third, within the high-compatibility condition, approximately equal numbers of participants were categorized as either responding predominantly color-first (7 out of 19), predominantly word-first (7 out of 19), or "mixed" (5 out of 19). Even among the 5 "mixed" participants, their choice of one response order or the other was fairly consistent *within each block*.

Given that the double-response Stroop data enable a categorization of participants according to their chosen response order, it is possible to reanalyze participants' Stroop interference

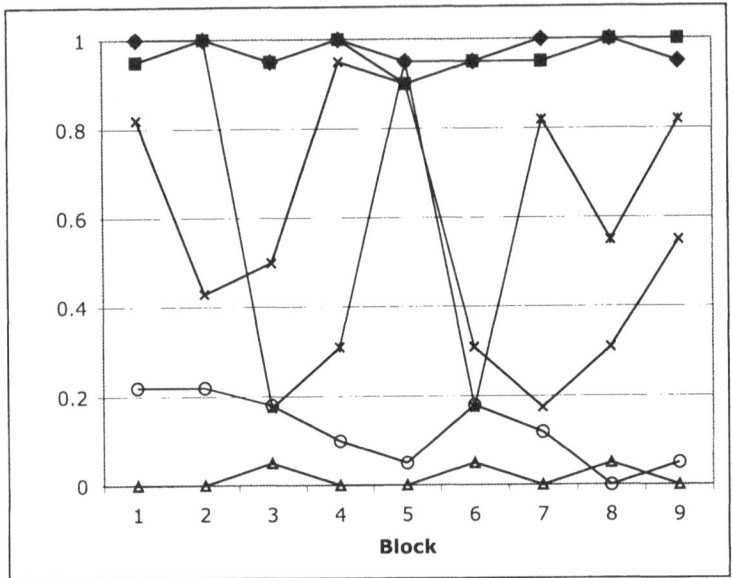

Fig. 4. Selected participants' response-order proportions from Experiment 1, presented by block, in the double-response task. Note that two of the participants predominantly responded word first (filled shapes), two predominantly responded color first (open shapes), and two showed mixed response orders.

effects separately for these groups. Interestingly, there is *no* reliable difference among the response-order groups in terms of the size of their *double-response* Stroop interference, $F(2, 16)$ = 2.4, mean square error [*MSE*] = 1113. This may reflect some degree of adaptivity in participants' response-order choices: They use a response order that meshes well with their skills and hence, any differences in Stroop interference are attenuated.

What is more interesting is that there is a difference among the response-order groups in terms of their standard Stroop interference. Focusing on the participants in the high-compatibility condition (because there is greater strategic variability in this group), participants who choose to respond to the color first in the double-response blocks show the least Stroop interference in the standard blocks (less than 0 msec on average). Participants who chose to respond to the word first in the double-response task show the greatest Stroop interference in the standard Stroop blocks (44 msec), and participants in the mixed group are intermediate (26 msec). In other words, participants who choose a double-response strategy that involves reporting the color first may be shielding themselves from interference in the standard Stroop task. Indeed, this group's interference effect is significantly smaller than the other two groups', $F(1, 16) = 4.0$, $MSE = 4580$, an effect that will be further explored in Experiment 2.

It should be emphasized that participants are self-selecting into these response-order groups, that is, this is not an experimental manipulation. Therefore, these results cannot distinguish between the following two interpretations. One interpretation is that the double-response task acts as a sensitive diagnostic regarding the variability in people's natural approach to Stroop tasks, that is, the degree to which they emphasize color in their processing and responding. Under this interpretation, the difference in Stroop interference across response-order groups simply reflects individual variation that is a common cause of Stroop interference magnitude and response-order choice. Consistent with this interpretation is the fact that color-only reaction time correlates significantly with proportion word-first responses (among the high-compatibility condition where there is sufficient response-order variability), $r = .323$, $p < .05$.

The second interpretation is that the double-response task acts as an effective manipulation to change participants' relative emphasis on the two stimulus dimensions, which they can then *transfer to the* standard Stroop task, hence impacting Stroop interference. For example, by practicing color-first responding in the double-response task, one is becoming more likely to emphasize the ink color and thus will be able to perform the standard task with less interference.

Note that the model presented in this article is consistent with *both* interpretations. If participants are tending toward a particular response order for the double-response task, they will show the same bias in processing for the standard task, which will be reflected in their standard Stroop interference. The model would account for these related effects by positing individual differences in participants' a priori utility values for color naming and word reading. If, on the other hand, participants are more randomly choosing (and maintaining) a response order, the practice they get in the double-response task will impact their word-reading and color-naming production rule utilities in such a way they will be biased toward the same order of processing for the standard Stroop task. Regardless of which interpretation holds, the model specifically predicts the latter effect, namely that response order in the double-response task will impact standard Stroop interference. This prediction is tested in the next experiment.

4. Experiment 2

The goal of Experiment 2 is to use an experimental manipulation of participants' response order in the double-response Stroop task to test whether the relation between response order and size of the standard task Stroop effect, found in Experiment 1, is a causal one. That is, does practicing the double-response Stroop task in one order versus the other differentially impact the Stroop effect (as measured in the standard task)?

The current model predicts that the color-first order will significantly reduce the size of participants' Stroop effect and that the word-first order will moderately increase the size of participants' Stroop effect. This difference is predicted by this model because of the utility-based learning it employs. Note that a practice-based model would not predict a difference because either double-response order involves the same amount of practice. Under a utility-learning model, the assumption is that participants come into the experiment with a higher utility associated with word reading than color naming, thus tending toward a prepotent word-reading response. If participants bring to the experiment a bias toward this word-first strategy and yet are assigned to the color-first condition, word-reading will incur a very high cost as participants essentially engage a word-then-color-then-word sequence of processing, where the first attempt at word processing is something of a false start. That extra cost will negatively impact the word-reading production rule's utility over the course of double-response practice, making word reading less likely to fire first, and hence producing smaller Stroop interference in subsequent standard Stroop blocks. In contrast, for participants assigned to the word-first condition, the initial tendency to respond word-first will be rewarded, that is, for them, reading the word first is an immediate success. This may lead to an increase in the utility of word reading, but because it is already quite high relative to color naming, the model does not predict as large an increase in Stroop interference in this condition as it does a reduction in the other, color-first condition.

4.1. Methods

4.1.1. Participants

The participants in this experiment were 41 Carnegie Mellon University undergraduates who received course-related credit for participating. One additional participant was not included in the analysis for failure to follow instructions.

4.1.2. Design

The design of this experiment includes one between-subject factor called *order*, which specifies the order in which participants were instructed to respond to the two stimulus dimensions in the double-response Stroop blocks—either color then word or word then color. In this experiment, all participants used the same response modalities as the high-compatibility condition from Experiment 1: saying the word and keying in the color. Thus, color-then-word participants were always keying-then-saying their responses, and word-then-color participants were always saying then keying.

The within-subjects factors in this experiment correspond to the different blocks participants completed—all in a fixed order: word-only, word-only, color-only, color-only, standard Stroop, standard Stroop, double-response (eight contiguous blocks), standard Stroop, standard Stroop, color-only, word-only. This design enables us to measure changes in performance (particularly, standard Stroop interference) from the pre-double-response blocks to the post-double-response blocks.

Each standard and double-response block had an equal number of conflict and congruent trials randomly permuted within a 24-trial block.

4.1.3. Procedure and apparatus

The procedure and apparatus were the same as in Experiment 1 except that the experimental software would not continue with the next trial until a response of each type had been given in the correct order.

4.1.4. Results and discussion

This section is divided into three parts corresponding to each of three phases of the experiment: before, during, and after the double-response blocks in which the experimental manipulation occurred. Note that the key model prediction involves a change in Stroop interference across the experiment; this is tested in the subsection on performance *after* the double-response blocks.

4.1.4.1. Performance before the double-response blocks. Figure 5a displays the reaction time and error data for both experimental groups for the three block types occurring before the double-response blocks. Participants in both groups were fast and highly accurate in respond-

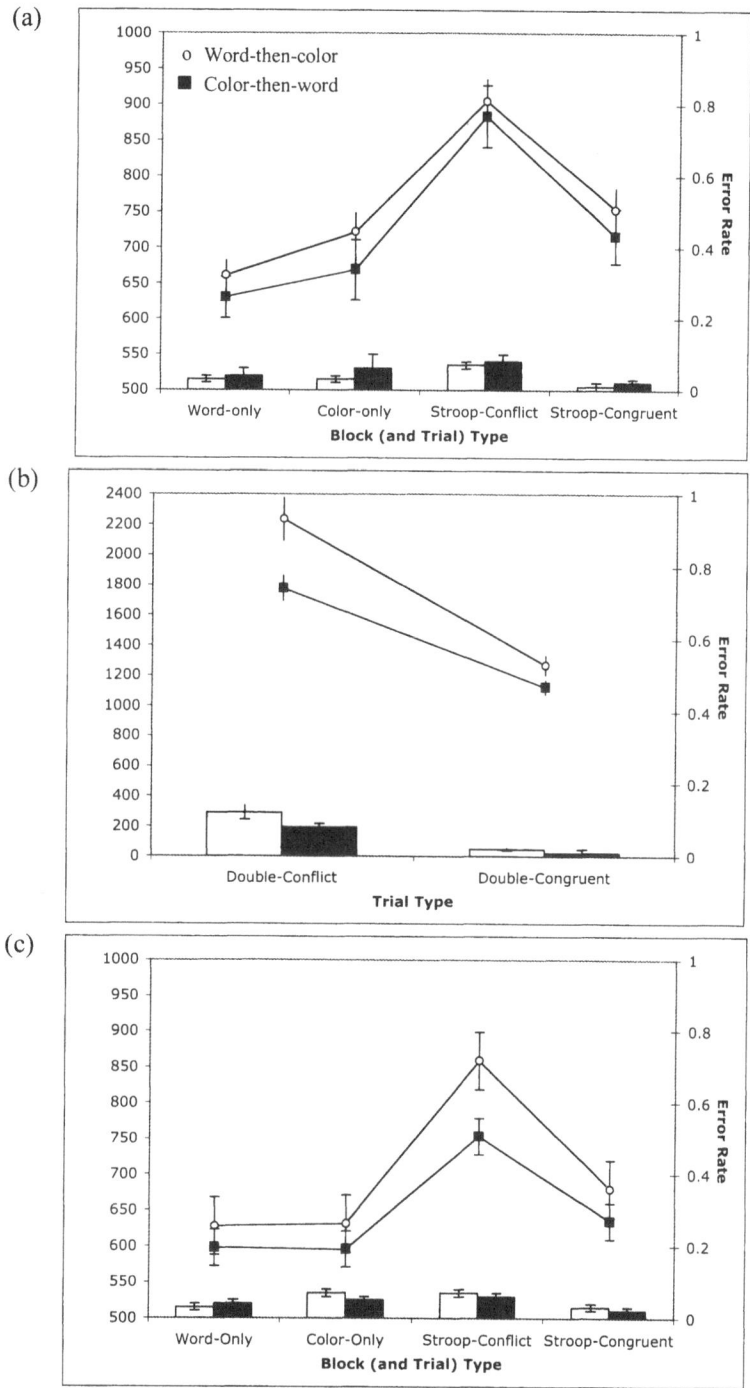

Fig. 5. Experiment 2 reaction times (lines) and error rates (bars) *before* the double-response blocks (panel a), *during* the double-response blocks (panel b), and *after* the double-response blocks (panel c). The left vertical axis corresponds to the reaction-time data, and the right vertical axis corresponds to the error rate data. Open circles and open bars refer to word-then-color condition; filled squares and filled bars refer to color-then-word condition.

Table 3
ANOVA results for Experiment 2, pre-double-response blocks

Task: Dependent Measure	Factor Analyzed	F	MSE	p value
Word-only: RT	Group	$F(1, 39) = .77$	12700	*ns*
Word-only: Errors	Group	$F(1, 39) = .35$	.003	*ns*
Color-only: RT	Group	$F(1, 39) = 1.15$	24200	*ns*
Color-only: Errors	Group	$F(1, 39) = .80$	.014	*ns*
Standard Stroop: RT	Group	$F(1, 39) = .37$	44400	*ns*
	Trial type	$\mathbf{F(1, 39) = 65}$	8050	**.0001**
	Group × Trial type	$F(1, 39) = .17$	8050	*ns*
Standard Stroop: Errors	Group	$F(1, 39) = 1.0$	.003	*ns*
	Trial type	$\mathbf{F(1, 39) = 29.5}$	.002	**.0001**
	Group × Trial type	$F(1, 39) = .02$	.002	*ns*

Note. Boldface indicates significant effect. ANOVA = analysis of variance; *MSE* = mean square error; RT = reaction time; *ns* = nonsignificant.

ing to word-only trials, relatively fast and highly accurate in responding to color-only trials, and showed significant Stroop interference in both reaction time (approximately 150 msec) and error rate (6%) for standard Stroop trials. This is consistent with other reports of Stroop interference.

For the pre-double-response blocks, performance between the two experimental conditions did not differ; that is, random assignment to the two conditions was satisfactory. See Table 3 for F values of each ANOVA, conducted as a one-way analysis (with experimental group as the single factor) for the word-only and color-only tasks and as a two-way mixed analysis (Experimental group × Trial type) for the standard Stroop task. Of all these analyses, the only significant effect was the main effect of trial type (i.e., significant Stroop interference) for both the reaction time and error data in the standard Stroop blocks.

4.1.4.2. Performance during the double-response blocks. Figure 5b shows total reaction times and error rates for both experimental groups for the double-response Stroop trials. Note that, for the double-response task, a trial is counted as having an error if *either* response was incorrect. These data were submitted to a two-way, mixed ANOVA, with response order as the between-subject factor and trial type as the within-subjects factor. Results indicated that the word-then-color group took longer, $F(1, 39) = 6.9$, $p < .05$, and made more errors, $F(1, 39) = 6.0$, $MSE = .002$, $p < .05$, than the color-then-word group. Both groups showed a significant effect of trial type—that is, significant Stroop interference—in reaction times, $F(1, 39) = 204$, $p < .001$, and errors, $F(1, 39) = 97.1$, $MSE = 4.0$, $p < .001$. Moreover, the Group × Trial-type interaction showed that the size of Stroop interference was larger for the word-then-color group than for the color-then-word group for reaction times, $F(1, 39) = 7.6$, $p < .01$, and marginally for errors, $F(1, 39) = 4.03$, $MSE = .001$, $p = .05$.

These between-group differences highlight the difficulty of the word-then-color response order and argue against different speed-accuracy trade-offs between groups. Finally, it should be noted that, for both groups, total reaction times for double-response conflict trials were more than twice those for standard Stroop conflict trials.

4.1.4.3. Performance after the double-response blocks. Figure 5c presents the reaction time and error data for both experimental groups for the three postmanipulation block types. Recall that this model's prediction is that practicing color-then-word response order in the double-response trials will increase the utility associated with color-processing production rules enough to favor processing the color dimension. This strategy shift toward color-naming first will, in turn, reduce Stroop interference in the subsequent standard Stroop blocks. Conversely, practicing word-then-color response order in the double-response trials will accentuate participants' preexisting high utility associated with word-processing production rules and hence maintain or slightly increase Stroop interference in the subsequent standard Stroop blocks.

To test these hypotheses regarding changes in Stroop interference, the data from the standard Stroop blocks occurring before and after the double-response blocks were submitted to a $2 \times 2 \times 2$ mixed ANOVA, with factors being group (color-then-word vs. word-then-color), trial type (conflict vs. congruent), and phase (pre- vs. postmanipulation). For the error data, there was only one significant effect, namely that more errors were committed on conflict trials, $F(1, 39) = 31.1$, $MSE = .003$, $p < .001$.

There were two main effects and one interaction in the reaction-time data. First, participants' reaction times were faster in the postmanipulation phase than in the premanipulation phase, $F(1, 39) = 21.4$, $MSE = 13000$, $p < .001$. This is consistent with a general speedup from increased familiarity with the experimental setup and with the key mapping for the colors. Second, participants were slower on conflict trials, $F(1, 39) = 95$, $MSE = 10280$, $p < .001$. This result points again to significant Stroop interference. The 3-way interaction of Group × Trial type × Phase reached marginal significance, $F(1, 39) = 3.2$, $MSE = 4750$, $p = .08$. This interaction is consistent with the hypotheses stated earlier about *different* changes in Stroop interference for the two groups. To test whether this interaction is of the form predicted by our hypotheses, we computed each participant's change in Stroop interference between the premanipulation and postmanipulation phases. That is, we created a new variable: (postStroopRT[conflict]—postStroopRT[congruent])—(preStroopRT[conflict]—preStroopRT[congruent]). Specifically, our prediction is to see a reduction in Stroop interference for the color-then-word group and a slight increase in (or maintenance of) Stroop interference for the word-then-color group. The data fit these predictions precisely: The color-then-word group on average showed a 50 (±27) msec reduction in Stroop interference, and the word-then-color group showed a 27 (±33) msec increase in Stroop interference. A two-sample *t* test comparing the conditions' change in Stroop interference showed a significant difference in the predicted direction, $t(39) = -1.8$, $p < .05$ (one-sided). In addition, comparing each group's average change in Stroop interference to zero revealed that the color-then-word group's change showed a significant reduction $t(20) = -1.8$, $p < 05$ (one-sided), and the word-then-color group's change was not significantly different from zero $t(19) = 0.814$.

Although the double-response response-order manipulation was predicted to impact Stroop interference, it was not predicted to impact the color-only or word-only blocks where there is no competing second dimension. Thus, the color-only and word-only blocks serve as control tasks. ANOVAs were performed on both block types' reaction time and error data, with group (word-then-color vs. color-then-word) and experimental phase (pre-double-blocks vs. post-double-blocks) as factors. In all cases the interaction was not significant, as expected. For reaction times, both block types showed significant speedup from pre- to postmanipulation:

$F(1, 39) = 30.7, MSE = 4500, p < .001$, for color-only blocks and $F(1, 39) = 12.3, MSE = 1700$, $p < .005$, for word-only blocks. This is likely the same generic speedup that we found for the Stroop blocks. It is somewhat surprising that there would be speedup in the word-only blocks because reading is so highly practiced, but the speedup was only on the order of 30 msec. For the error data in the color-only and word-only blocks, there were no significant effects.

4.1.5. Modeling change in Stroop interference

The same model used to produce the model fits discussed in the introduction was used to fit these data. The only difference is that for this experiment the model needed one extra parameter to initialize utility learning. Thus, with three free parameters, the model was fit to 18 data points, minimizing the mean deviation between model and data. Twelve of these data points were raw Stroop reaction times (see Figure 6), and the other six were conflict—congruent reaction-time differences, included to emphasize interference effects in optimizing the model fit. This model fit had a mean deviation of 62.1 msec and $R^2 = .97$.

Figure 6 shows that the model is able to capture several effects in the data: Conflict trials are slower than congruent trials, standard Stroop trials are faster than double-response trials, double-response word-first responses are slower than double-response color-first responses (although the model underestimates this effect), and response order in the double-response blocks impacts standard Stroop interference. This last result is the key finding of Experiment 2, namely that the size of Stroop interference decreased for the color-then-word group and (slightly) increased for the word-then-color group. The model similarly showed an initial Stroop interference effect of 176 msec, which decreased to 105 msec in the color-then-word condition and increased to 214 msec in the word-then-color condition.

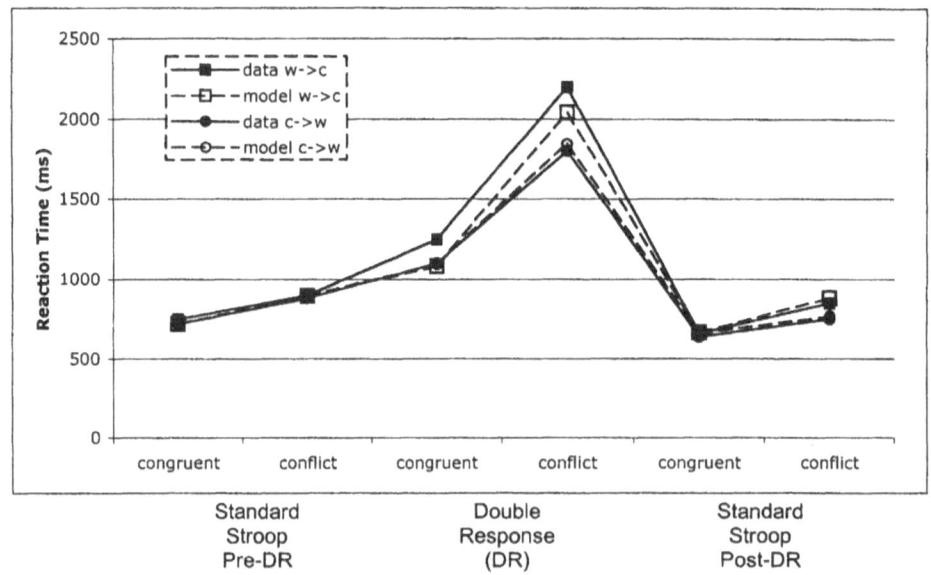

Fig. 6. Model fit to key trial types in Experiment 2.

These model results stem from the model's ability to initiate processing with either the word or color dimension and from its ability to shift its bias from one response order to another, based on their learned utilities. For example, because of its high initial utility for word reading, the model tends to process the "correct" stimulus dimension first in the word-then-color condition but the "wrong" one in the color-then-word condition. This experience updates the utility of production rules for processing each stimulus dimension, with more utility change occurring in the color-then-word condition. This bias carries forward to the post-double-response standard Stroop blocks, producing reduced Stroop interference in the color-then-word case and slightly increased Stroop interference in the word-then-color case.

5. Experiment 3

Experiment 2 shows that instructions specifying a particular response order in the double-response Stroop task can experimentally impact the size of participants' Stroop interference effect. Experiment 3 aimed to replicate this finding and expand it to a slightly different situation. The only procedural change from Experiment 2 to 3 was that all participants were asked to produce *both* responses manually. That is, the mapping of stimulus dimension (word vs. color) to response was not a mapping to modality (verbal vs. manual) but rather a mapping to a particular row of keys. This had the advantage of making the two responses more comparable in response times (i.e., simple horse-race accounts would be insufficient). Further, this offered a different approach to the response-mapping compatibility issues that surfaced in Experiment 1: Instead of eliminating these issues (as in Experiment 2 by studying only the high-compatibility response mapping), Experiment 3 made both responses manual. In effect, this should make the task somewhat more difficult than the compatible response mapping (Experiments 1 and 2) but less difficult than the incompatible response mapping (Experiment 1). Recall that this model predicts that an increase in task difficulty, especially when it increases response times, would lead to a larger change in Stroop interference after the double-response manipulation. Thus, the qualitative prediction going into this experiment is the same as Experiment 2: The color-then-word condition should show a reduction in Stroop interference after double-response blocks. Moreover, because of Experiment 3's more difficult version of the double-response Stroop task, this should be somewhat magnified relative to Experiment 2.

The only study-design change in Experiment 3 is that the second of the two experimental conditions was no longer another version of the double-response task but rather an equal number of trials of standard Stroop. This condition offers a new control for the color-then-word double-response condition. A reasonable expectation would be that more practice at standard Stroop in the manipulation phase would be the best way to reduce Stroop interference—even better than double-response Stroop. This model, in contrast, predicts reduction in Stroop interference for both groups, but more for the color-then-word double-response condition.

5.1. Methods

5.1.1. Participants

The participants in this experiment were 53 Carnegie Mellon University undergraduates who received course-related credit for participating.

5.1.1.1. Design.
The design of this experiment included one between-subject manipulation, specifying which task participants performed in the middle blocks of the experiment: either double-response Stroop (in the color-then-word order) or standard Stroop.

5.1.2. Procedure, apparatus, and analysis

Except for the differences noted previously, the methods were the same as in Experiment 2. Data analyses focus on reaction times and use the same analysis procedure as Experiments 1 and 2.

5.1.3. Results and discussion

This section is divided into three parts corresponding to the three phases: before, during, and after the experimental manipulation. Note that, as with Experiment 2, the critical model prediction involves a change in Stroop interference from pre- to postmanipulation blocks; this is tested in the section on performance *after* the manipulation.

5.1.3.1. Performance before the experimental manipulation.
Figure 7a displays reaction times for both groups for the three premanipulation block types: word-only, color-only, and standard Stroop. In the case of the standard Stroop blocks, reaction times are plotted separately for conflict and congruent trials. Participants in both groups took approximately 650 msec for the word-only and color-only tasks, took longer for the standard Stroop task, and showed a significant Stroop interference effect (approximately 150 msec), $F(1, 51) = 100.1$, $MSE = 6528$. This last finding is consistent with Stroop interference effects found in other reports, including Experiment 2. The word-only reaction time may seem longer than typical results in the literature, but recall that in this experiment participants were learning new key mappings for manual responding. Most other reports of word-only reaction times involve highly practiced verbal responses. Also note that, as in Experiment 2, the expectation was that performance in these premanipulation blocks should not differ between conditions. This expectation was supported for all three tasks, word-only: $F(1, 51) = 1.80$, $MSE = 17100$; color-only: $F(1, 51) = 0.99$, $MSE = 13690$; standard Stroop: $F(1, 51) = 2.2$, $MSE = 42400$.

5.1.3.2. Performance during the experimental manipulation.
During the experimental manipulation blocks, participants were performing either the double-response Stroop (responding to the color *first*) or the standard Stroop task (i.e., responding to the color *only*). Therefore, it is not surprising that the double-response group had much longer reaction times, $F(1, 51) = 47.5$, $MSE = 341280$. It is also not surprising that there was a significant trial-type effect, with conflict trials taking longer than congruent trials, $F(1, 51) = 43.2$, $MSE = 87300$. Similar to the result in Experiment 2, the interaction of these two effects was also significant, with the larger trial-type effect exhibited by the double-response group, $F(1, 51) = 27.9$, $MSE = 87300$. Again,

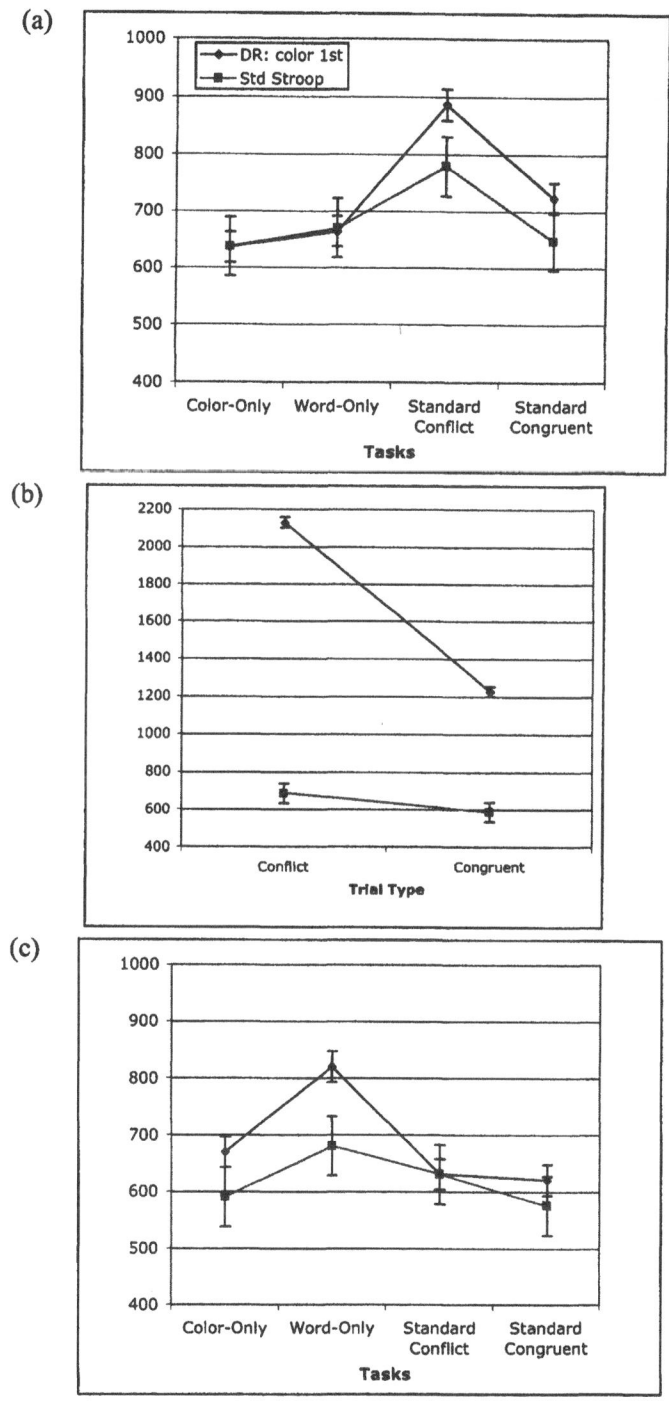

Fig. 7. Experiment 3 reaction-time data (a) *before* the experimental manipulation, (b) *during* the experimental manipulation, and (c) *after* the experimental manipulation.

this interaction is likely an example of Stroop interference scaling with the absolute magnitude of reaction times. In fact, looking at the data in Figure 7b, it is important to notice the scale of the y axis to realize that the difference between conflict and congruent trials for the standard Stroop participants is actually a full 100 msec; this effect appears small mainly because of the much larger double-response reaction times.

5.1.3.3. Performance after the experimental manipulation. Figure 7c presents the reaction times for both groups for the three postmanipulation block types: word-only, color-only, and standard Stroop. As in Experiment 2, the key model prediction involves testing whether the experimental manipulation had an impact on Stroop interference. In Experiment 2, that manipulation involved different response orders for the double-response Stroop, and the finding was that the color-first response order significantly reduced Stroop interference, whereas the word-first response order slightly amplified Stroop interference. In Experiment 3, the color-first double-response Stroop task was again expected to reduce Stroop interference. Here, however, the comparison condition was the standard Stroop task, another task where processing color first has increased utility. Thus, the model predictions for this experiment were that (a) both tasks will lead to a reduction in Stroop interference, and (b) this reduction will be greater for the double-response condition. These predictions were tested by a 2×2 mixed ANOVA on the size of Stroop interference, where the between-subject factor was the experimental manipulation (double-response or standard Stroop) and the within-subjects factor was phase of the experiment (pre- or postmanipulation). Supporting prediction (a), this analysis showed a significant main effect of phase of the experiment (pre- vs. postmanipulation), such that Stroop interference effects were smaller after either task manipulation, $F(1, 51) = 26.9$, $MSE = 7276$. Supporting prediction (b), these data showed an interaction between condition and phase of the experiment. As predicted, greater reduction in Stroop interference occurred for the double-response group, $t(51) = 1.77, p < .05$. (Note that this analysis is equivalent to the *t* test computed in Experiment 2, comparing pre- to postmanipulation changes in Stroop interference between conditions.)

Analyzing participants' reaction times in the word-only and color-only blocks before versus after the manipulation, only two significant effects were observed. The first of these was a significant increase in word-only reaction times, $F(1, 50)$, 14.2, $MSE = 9684$. This main effect, however, should be interpreted in light of its interaction with condition, $F(1, 50) = 5.6$, $MSE = 9684$. Specifically, this interaction showed that the double-response group showed a greater increase in word-only reaction times. These results are consistent with the utility of word reading decreasing relative to color naming for both conditions but more so in the color-then-word double-response case.

5.1.4. Modeling change in Stroop interference

The model that was fit to Experiment 2 was slightly adjusted to accommodate the main procedural change in Experiment 3, namely, that all participants were responding manually to both stimulus dimensions. This change was incorporated into the model by no longer considering all the word-reading association chunks to have high activation. In fact, the word-association chunks were set to have the same activation as the color-association chunks, and this activation value was a free parameter in the model fit. Although this activation value was adjusted to account for the new (manual) response mapping, the production rules for word processing and color processing started the experiment as they had in Experiment 2, with a utility-based bias toward processing the word first. From this starting point, the utility values of the two production rules shifted, based on the experience participants received in the two conditions. For the double-response condition, the utility values shifted as they had for the corresponding (color-then-word) condition in Experiment 2. For the standard Stroop condition, a single parameter was applied to account for the utility learning among participants in that group. All other parameters of the model were kept the same as the fit from Experiment 2.

Figure 8 shows the model fit for both conditions, for conflict and congruent trials across the three phases of the experiment—premanipulation standard Stroop trials, manipulation trials (either double-response or standard Stroop), and postmanipulation standard Stroop trials. This fit has $R^2 = .99$ and a mean deviation of 43.8 msec. Notice the data for the two groups in the premanipulation phase are numerically different. Although this is not a statistically reliable difference, the sampling effect of slightly faster reaction times among those in the standard Stroop condition is not accounted for by the model.

The model captures the four key results here. First, it captures the approximately 170 msec premanipulation Stroop effect as it had in Experiment 2. Second, it captures the very large Stroop effect in the double-response task as it had in Experiment 2. Third, it captures a gradual shrinking of the Stroop effect for the standard Stroop condition across the experiment by its shifting of the utility values for reading and color naming. Fourth, it captures a more substantial reduction in the postmanipulation Stroop effect for the double-response condition because of a more substantial shift in utility values. As discussed earlier, the utility value for color naming relative to word reading shifts more in the double-response condition because there is a greater benefit to processing the color first in this condition.

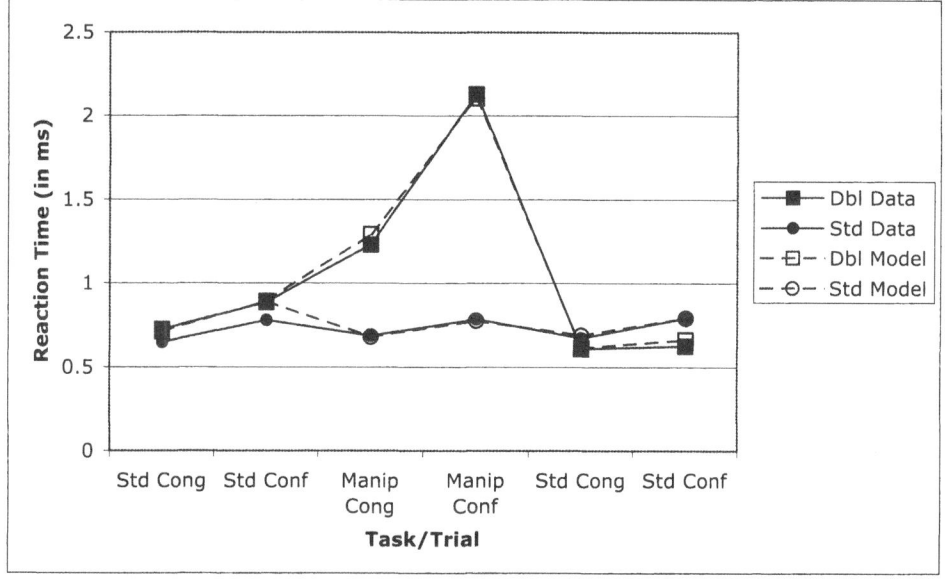

Fig. 8. Model fit to key trial types in Experiment 3.

6. General discussion

This article presented new modeling and empirical work regarding Stroop effects. The main idea behind this model is that people can approach Stroop tasks by processing the color *or* the word first and that this implicit choice reflects a learned (although likely still implicit), strategy choice based on the relative utility of word reading versus color naming. A model implementing this idea within the ACT–R architecture accounts for basic Stroop effects and several new Stroop results.

Specifically, this model made several novel predictions regarding particular Stroop-related experiences that could change the size of participants' Stroop interference. First, the model predicted a significant *decrease* in standard Stroop interference after double-response Stroop trials under color-then-word response-ordering instructions. Second, the model predicted a moderate *increase* in standard Stroop interference after double-response Stroop trials under word-then-color instructions. These predictions were confirmed in an observational study (Experiment 1) and under an experimental manipulation (Experiment 2). Third, the model predicted that additional practice at the standard Stroop task would also reduce the size of Stroop interference but not as much as experience at responding color-then-word in the double-response task. This prediction was also supported (Experiment 3).

6.1. Model implications

Utility learning is the key mechanism that this model uses to explain the observed Stroop effects. In this model, utility is a measure of how effective a production rule has been, averaged over the instances in which it was applied. Specifically, utility is computed in terms of successes, failures, and costs. This emphasis on the outcome of a production rule's firing over its sheer amount of practice harkens to the Law of Effect versus the Law of Practice (Thorndike, 1927, 1932) The Law of Effect states that actions that are effective (e.g., in achieving goals) will be repeated in the future. This is what happens with the color-naming production rule in some of the conditions investigated here, and when it is particularly effective, it becomes more likely to fire at the beginning of a trial, thus reducing subsequent Stroop interference. The insight of this model is that utility, a construct akin to effectiveness, is learned at the level of the production rule and that *relative* utility is what determines repeated future use. This is consistent with the results presented in this article and with past work (Lovett & Anderson, 1996; Singley & Anderson, 1989).

In contrast, the Law of Practice states that mere practice of an action leads to its repeated use. This view is more consistent with practice-based models of Stroop (e.g., Cohen et al., 1990). It predicts that practicing a skill—regardless of the effectiveness of that skill—will lead to more use of the skill. This is not what was found in these studies. For example, the two response orders for the double-response Stroop task had the same amount of practice with color naming and word reading but produced different changes—a decrease versus an increase—in Stroop interference depending on the response order. Practice-based models would predict a reduction in interference under both conditions.

This model includes both learning based on effectiveness and learning through practice. As an illustration of how it differs from a practice-based model, we fitted a practice-only variant of this model to the data from Experiment 2. Specifically, we disabled this model's utility-learning features but retaining its practice-based features, that is, the strengthening of color-association and word-association chunks. This is a reasonable practice-based model and is very similar to earlier ACT–R models of Stroop. Although this model produces an overall decrease in reaction times with practice that is consistent with the data, even the best-fitting version of this model fails to produce different-sized reductions in Stroop interference for the two response-order conditions: Its initial Stroop interference of 180 msec decreases to 164 msec for both conditions. The fit of this practice-only model (with mean deviation = 116.6 msec, R^2 = .94) is significantly worse than the fit of this model presented earlier ($p < .05$). This failure to adequately account for changes in Stroop interference is also consistent with results from practice-based models in the literature. For example, Figure 12 in Cohen et al. (1990) shows model performance that decreases its reaction times with practice, but that model produces changes in Stroop interference that differ systematically from what is observed in the MacLeod and Dunbar (1988) study. In contrast, this model was able to fit the changing nature of these Stroop interference data quite well (see Figure 2).

Botvinick et al. (2001) extended the practice-based model of Cohen et al. (1990) by adding another layer that can adjust its task-related control. This model posits that greater control (from task control nodes) is exerted when there is more "crosstalk" in the system, where crosstalk is a model-based measure of conflict in the network. With this extra component, the model can account for data from Tzelgov et al. (1992) that show a reduction in the size of Stroop interference as a function of proportion of conflict trials. As mentioned in Section 2.3, this model can also address this effect because the key production rules' utility values change as a result of experience (e.g., lower accuracies and slower responses with a greater proportion of conflict trials). In the Botvinick et al. model, control is modulated by crosstalk. As a rule of thumb, crosstalk tends to be high when performance is low, and low when performance is high. This is consequently a proxy for processing effectiveness, which we see as introducing an element of utility into a practice-based account.

Similarly, one could conceive of a practice-based model variant in which a greater learning benefit is gained during double-response Stroop from practice on the first-processed dimension than the second-processed dimension (e.g., due to some attentional bias). This account would be consistent with the results in Experiment 2 in that it would produce greater benefit for color processing in the color-then-word condition and hence would also produce subsequent reduced Stroop interference. But this account also predicts a greater reduction in Stroop interference after standard Stroop practice—where color processing would get full attention—compared to double-response practice. This is inconsistent with the results from Experiment 3.

6.2. Learning new production rules

One feature of this model that plays a role in the predictions discussed throughout this article is that it uses many of the same production rules in the double-response and standard Stroop tasks. That enables the utility-learning effect gained from the double-response trials to carry over to strategy choices in the standard Stroop task. Employing the same production rules in both tasks is parsimonious in that the double-response task is simply a combination of standard

Stroop tasks. Moreover, the model's fit to the data in Experiments 2 and 3 supports this knowledge representation. An interesting question, however, is whether people could learn a new production rule for the double-response task given sufficient practice. If so, any utility learning in the double-response task would apply to this new production and would not carry over to the standard Stroop task. This idea suggests that giving participants more and more experience at the double-response Stroop task could eventually lead to a *reduced* impact on their standard Stroop interference effect. This idea has been explored in Anderson et al. (2004) and is at the heart of empirical questions currently being explored by Hazeltine and colleagues (Hazeltine, Teague, & Ivry, 2002).

What can we say about reducing Stroop interference?

Because the Stroop effect is so robust, any result that shows it being reduced or eliminated warrants attention. Several manipulations have been shown to reduce Stroop interference, and some of these have been highlighted in this article. First, most simply, additional practice at the standard Stroop task does reduce Stroop interference, albeit gradually. Experiment 3 of this article showed a reduction in Stroop interference from approximately 160 msec down to 100 msec. This occurred with standard Stroop trials distributed equally between conflict and congruent types but where word *and* color responses were made manually. Note that the verbal-response Stroop would produce a smaller reduction in interference because verbal responses are faster than manual responses and thus leave less room for utility changes. Other results that show changes in Stroop interference were discussed earlier. These include the uneven distribution of trial types (Stroop, 1935; Tzelgov et al., 1992) and the learning of a new competing task (MacLeod & Dunbar, 1988).

It should be emphasized that these approaches to reducing Stroop interference require more than simply giving more standard Stroop practice. In each case, there was something special about the Stroop task participants performed: unusual response mode, biased trial-type composition, or novelty of the tasks involved. A reasonable heuristic seems to be that the less "standard" the Stroop task being practiced, the greater impact it will have on reducing Stroop interference. From the model's point of view, this is explained by the fact that standard Stroop task practice does not allow for a very large utility shift between word reading and color naming. Task variants that place a higher cost on word reading (or conversely, a lower cost on color naming) will show the largest change in utility and hence the greatest possible reduction in Stroop interference.

References

Altmann, E. M., & Davidson, D. J. (2001). An integrative approach to Stroop: Combining a language model and a unified cognitive theory. In J. D. Moore & K. Stenning (Eds.), *Proceedings of the 23rd annual conference of the Cognitive Science Society* (pp. 21–26). Hillsdale, NJ: Lawrence Erlbaum Associates, Inc.

Anderson, J. R., Bothell, D., Byrne, M. D., Douglass, S., Lebiere, C., & Qin, Y. (2004). An integrated theory of the mind. *Psychological Review, 111,* 1036–1060.

Anderson, J. R., & Lebiere, C. (1998). *The atomic components of thought.* Mahwah, NJ: Lawrence Erlbaum Associates, Inc.

Botvinick, M. M., Braver, T. S., Barch, D. M., Carter, C. S., & Cohen, J. D. (2001). Conflict monitoring and cognitive control. *Psychological Review, 108,* 624–652.

Cheesman, J., & Merikle, P. M. (1984). Priming with and without awareness. *Perception and Psychophysics, 36,* 387–395.

Chrisman, S. D. (2001). Individual differences in Stroop and local-global processing: A possible role of interhemispheric interaction. *Brain & Cognition, 45,* 97–118.

Cohen, J. D., Dunbar, K., & McClelland, J. L. (1990). On the control of automatic processes: A parallel distributed processing account of the Stroop effect. *Psychological Review, 97,* 332–361.

Cohen, J. D., & Huston, T. A. (1994). Progress in the use of interactive models for understanding attention and performance. In C. Umilta, & M. Moscovitch (Eds.), *Attention and performance 15: Conscious and nonconscious information processing* (pp. 453–476). Cambridge, MA: MIT Press.

Comalli, P. E., Jr., Wapner, S., & Werner, H. (1962). Interference effects of Stroop color-word test in childhood, adulthood, and aging. *Journal of Genetic Psychology, 100,* 47–53.

Dalrymple-Alford, E. C., & Azkoul, J. (1972). The locus of interference in the Stroop and related tasks. *Perception & Psychophysics, 11,* 385–388.

Dalrymple-Alford, E. C., & Budayr, B. (1966). Examination of some aspects of the Stroop color-word test. *Perceptual & Motor Skills, 23*(3, Pt. 2), 1211–1214.

Glaser, M. O., & Glaser, W. R. (1982). Time course analysis of the Stroop phenomenon. *Journal of Experimental Psychology: Human Perception and Performance, 8,* 875–894.

Greenwald, A. G. (1972). On doing two things at once: Time sharing as a function of high ideomotor compatibility. *Journal of Experimental Psychology, 94,* 52–57.

Hazeltine, E., Teague, D., & Ivry, R. (2002). Simultaneous dual-task performance reveals parallel response selection after practice. *Journal of Experimental Psychology: Human Perception and Performance, 28,* 527–545.

Kane, M., & Engle, R. W. (2003). Working-memory capacity and the control of attention: The contributions of goal neglect, response competition, and task set to Stroop interference. *Journal of Experimental Psychology: General 132,* 47–70.

Kerns, J. G., Cohen, J. D., MacDonald, A. W. III, Cho, R. Y., Stenger, V. A., & Carter, C. S. (2004). Anterior cingulate conflict monitoring and adjustments in control. *Science, 303,* 1023–1026.

Klein, G. S. (1964). Semantic power measured through the interference of words with color-naming. *American Journal of Psychology, 77,* 576–588.

Levelt, W. J. M., Roelofs, A., & Meyer, A. S. (1999). A theory of lexical access in speech production. *Behavioral and Brain Sciences, 22*(1), 1–38.

Lindsay, D. S., & Jacoby, L. L. (1994). Stroop process dissociations: The relationship between facilitation and interference. *Journal of Experimental Psychology: Human Perception and Performance, 20,* 219–234.

Logan, G. D. (1980). Attention and automaticity in Stroop and priming tasks: Theory and data. *Cognitive Psychology, 12,* 523–553.

Logan, G. D., & Zbrodoff, N. J. (1979). When it helps to be misled: Facilitative effects of increasing the frequency of conflicting stimuli in a Stroop-like task. *Memory and Cognition, 7,* 166–174.

Logan, G. D., & Zbrodoff, N. J. (1982). Constraints on strategy construction in a speeded discrimination task. *Journal of Experimental Psychology: Human Perception and Performance, 8,* 502–520.

Logan, G. D., Zbrodoff, N. J., & Williamson, J. (1984). Strategies in the color-word Stroop task. *Bulletin of the Psychonomic Society, 22,* 135–158.

Lovett, M. C. (1998). Choice. In J. R. Anderson, & C. Lebiere (Eds.), *The atomic components of thought* (pp. 255–296). Mahwah, NJ: Lawrence Erlbaum Associates, Inc.

Lovett, M. C. (2002). Modeling selective attention: Not just another model of Stroop. *Cognitive Systems Research, 3,* 67–76.

Lovett, M. C., & Anderson, J. R. (1996). History of success and current context in problem solving. *Cognitive Psychology, 31,* 168–217.

MacLeod, C. M. (1991). Half a century of research on the Stroop effect: An integrative review. *Psychological Bulletin, 109,* 163–203.

MacLeod, C. M. (1998). Training on integrated versus separated Stroop tasks: The progression of interference and facilitation. *Memory & Cognition, 26,* 201–211.

MacLeod, C. M., & Dunbar, K. (1988). Training and Stroop-like interference: Evidence for a continuum of automaticity. *Journal of Experimental Psychology: Learning, Memory, and Cognition, 15,* 126–135.

MacLeod, C. M., & MacDonald, P. A. (2000). Inter-dimensional interference in the Stroop effect: Uncovering the cognitive and neural anatomy of attention. *Trends in Cognitive Sciences, 4,* 383–391.

Neill, W. T. (1978). Decision processes in selective attention: Response priming in the Stroop color-word task. *Perception and Psychophysics, 23,* 80–84.

Newell, A. (1990). *Unified theories of cognition.* Cambridge, MA: Harvard University Press.

Phaf, R. H., van der Heijden, A. H., & Hudson, P. T. (1990). SLAM: A connectionist model for attention in visual selection tasks. *Cognitive Psychology, 22,* 273–341.

Roelofs, A. (2002). Control of language: A computational account of the Stroop asymmetry. In N. Taatgen & J. Aasman (Eds.), *Proceedings of the Third International Conference on Cognitive Modeling* (pp. 234–241). Veenendaal, The Netherlands: Universal Press.

Roelofs, A. (2003). Goal-referenced selection of verbal action: Modeling attentional control in the Stroop task. *Psychological Review, 110,* 88–125.

Roelofs, A., & Hagoort, P. (2002). Control of language use: Cognitive modeling of the hemodynamics of Stroop task performance. *Cognitive Brain Research, 15*(1), 85–97.

Schiller, P. H. (1966). Developmental study of color-word interference. *Journal of Experimental Psychology, 72,* 105–108.

Schooler, C., Neumann, E., Caplan, L. J., & Roberts, B. R. (1997). A time course analysis of Stroop interference and facilitation comparing normal individuals and individuals with schizophrenia. *Journal of Experimental Psychology: General, 126,* 19–36.

Schweickert, R. (1978). A critical path generalization of the additive factor method: Analysis of a Stroop task. *Journal of Mathematical Psychology, 18,* 105–139.

Shimada, H., & Nakajima, Y. (1991). Double response to Stroop stimuli. *Perceptual & Motor Skills, 72,* 571–574.

Siegler, R. S. (1991). Strategy choice and strategy discovery. *Learning and Instruction, 1*(1), 89–102.

Siegler, R. S. (1996). *Emerging minds.* New York: Oxford University Press.

Singley, M. K., & Anderson, J. R. (1989). *The transfer of cognitive skill.* Cambridge, MA: Harvard University Press.

Stroop, J. R. (1935). Studies of interference in serial verbal reactions. *Journal of Experimental Psychology, 18,* 643–662.

Sugg, M. J., & McDonald, J. E. (1994). Time course of inhibition in color-response and word-response versions of the Stroop task. *Journal of Experimental Psychology: Human Perception and Performance, 20,* 647–675.

Thorndike, E. L. (1927). The law of effect. *American Journal of Psychology, 39,* 212–222.

Thorndike, E. L. (1932). *The fundamentals of learning.* New York: Columbia University.

Tzelgov, J., Henik, A., & Berger, J. (1992). Controlling Stroop effects by manipulating expectations for color words. *Memory & Cognition, 20,* 727–735.

Virzi, R. A., & Egeth, H. E. (1985). Toward a translational model of Stroop interference. *Memory & Cognition, 13,* 304–319.